The Essentials of High School Math

Covering the Standards Set by the
National Council of Teachers of Mathematics

by Joseph R. Davis

Willow Tree Publishing

P.O. Box 302
Whitinsville, MA 01588

Comments, questions, praise, and/or criticism from
teachers *and* students are welcome at anytime at
comments@willowtreepublishing.com

When writing, feel free to write anonymously, but
please indicate the name of your school, city/town, and
state. All questions will be answered, and all
comments will be taken into consideration when
making changes to this book for future editions.

Any and all input is valued and appreciated.

The Essentials of High School Math

Copyright © Willow Tree Publishing. All rights reserved.
Cover Photo © 2009 JupiterImages Corporation. All rights reserved - used with permission.
Printed by Outskirts Press, Inc., Denver, Colorado

ISBN: 978-0-615-26509-4
Library of Congress Control Number: 2009920254

http://www.willowtreepublishing.com

Printed in the United States of America

Table of Contents

Unit 3 – Geometry and Measurement

Unit 4 – Statistics and Probability

About This Book

This book was designed to help students learn the basics of mathematics that they are *supposed* to understand upon entering high school, as well as the fundamental lessons within algebra, geometry, and statistics that students typically learn in ninth and tenth grade. It is intended for use in a semester-long remediation math course for students identified as possibly having trouble on an upcoming state-mandated math exam, or for those who have already failed such an exam and must re-take it.

There are sixty-one lessons in the book that are generally designed to be one-day lessons. Some may take two days, and a few may be combined with another lesson or other material, such as sample questions from your state's math exam, in order to fill an entire period of work. Since each state's high school math exam is different, providing sample test questions for students to practice is beyond the scope of this book. However, it is recommended that students attempt actual test questions from previous exams that should be available from your state Department of Education website. These test questions should be mixed in with the practice problems at the end of each lesson when time allows, especially at the beginning and end of each unit so students can see that their knowledge and test-taking skills have improved.

Since there are no official national math standards from the U.S. Department of Education, the lessons of this book are matched against the Standards for School Mathematics for Grades 6–8 by the National Council of Teachers of Mathematics (NCTM). (The second page of each unit shows how each lesson matches the learning expectations from the Standards, which are on pages 2, 74, 200, and 310 of the book.) This was done to show how this book covers nearly everything that any student should know mathematically before entering high school. There are additional lessons that build upon these Grade 6–8 standards in order to provide remediation on the algebra, geometry, and statistics basics that students typically learn up through tenth grade.

One important feature of this book is that it was created to include everything a student needs to study, practice, and learn the material that could show up on a state math exam. It contains almost 1000 problems for students to practice, and the lessons themselves contain about 100 questions from actual state exams that are thoroughly explained, along with hundreds of other examples. Each lesson is designed to cover one topic only so that students may thoroughly understand that topic.

In the interest of covering the greatest amount of material with the fewest number of questions, many state exams use questions that touch on several topics per question, such as a statistics question that requires interpreting a graph, or a geometry question that requires the solving of algebraic equations. While this may be a more efficient method for testing students, it can cause someone who understands most of a question to get it wrong. It also can provide misleading data about what skills students are lacking when test results come back to school districts. As such, while it's already been mentioned that students should be provided with questions from your state exam when possible, the practice questions in this book focus on the one topic that a lesson covers.

This book does **not** include any pre-made tests, quizzes, pre-tests, or post-tests. There are many different ways that a school might arrange to prepare students for their state math exam with an in-school or after-school tutoring/remediation class (or classes), and each specific method of preparation would likely require a different way of measuring student achievement. Whether students are quizzed or tested each week, after each unit, or even at all, and how this quizzing or testing is done, should be left up to each individual teacher to decide, given the circumstances of what is best for his or her class.

Taking a Standardized Math Test – Tips and Suggestions

When the time comes to take any state-mandated standardized math test, read through and attempt every question once before moving on. If you don't know how to answer a question, at least read the entire problem and possible answers (if it is a multiple-choice question) before going to the next question. By doing this, when you answer questions you **can** do, the ones you've read so far but skipped will be in the back of your mind, and it's possible to figure out questions you were stuck on while working through other problems you know how to do — almost like your brain warms up while you take the test. But this only works *if you read each question carefully* before moving on to the next one.

After going through the entire test (or entire session for that day), you should return to any open-response questions next. There may only be two or three open-response questions, but they are usually worth a significant percentage of all the possible points. And since most of the questions are multiple-choice, it can very tempting to answer only those and blow off the open-response questions without realizing how many possible points you are giving up. (Remember, multiple-choice and short-answer questions are typically worth one point each, but open-response questions can be worth more. On some state exams they are worth as much as four points each — and you can get *partial credit* on these.)

During a long (and possibly stressful) test, students can get tired of working and just want to get the test over with. This is why you should first make one complete pass through the test, and then go back to any open-response questions you skipped to make sure you've done all you can with *those* before reaching this feeling of being burned out by the exam. After your second attempts with any remaining open-response questions, then go back through the rest of the multiple-choice or short-answer questions and do your best to finish these — but always answer the open-response questions as much as possible! Even attempting to set up an open-response question you don't know how to do could get you one of the possible points (if the question is worth more than one point, as most open-response questions are).

When attempting multiple-choice questions, there are a couple things to remember: Of the four possible answers, two are usually way off from the correct answer, but the other wrong answer is usually very similar to the correct answer and may often be a solution that students find if they make a known common mistake. This means that the multiple-choice questions have two answers that might be easily ruled out, but it also means that each multiple-choice problem has a "trick" answer waiting that you might mistakenly choose.

The way to approach a multiple-choice question is as follows:

1) Try to work through the problem as if the four possible answers weren't there, and then if you can work to an answer, find which one of the four it is as a way of double checking your work. (Note: Remember, if you make a mistake while solving, you might come to an answer that is wrong but listed as one of the four choices, so be careful!)

2) If you aren't sure how to begin a problem, you can use the given answers to your advantage by trying to work backwards through the problem. Try out each answer in an equation, formula, or whatever is given for you to work with, and see if you can find the correct answer this way.

3) If you don't know how to solve the problem, or can't work backwards to a solution, next try to eliminate at least **two** of the possible answers. Sometimes when you can't find the exact solution, you can still recognize what answers are wrong. If you can eliminate *some* of the answers but can't go any further, then just guess at this point. You have a better chance of guessing the correct answer out of two choices than guessing one out of four.

4) If you can't work through the problem, or work backwards, or eliminate any of the answers, *don't leave the answer blank*! Still make a guess! A one out of four chance of getting a question correct is better than zero chance of getting it correct if you leave it blank. (Note: This only applies to state assessments, such as end-of-course or graduation exams. This advice does **not** apply to the PSAT or SAT where points are *deducted* for wrong answers. On the PSAT or SAT, if you could not eliminate any possible answers, do **not** guess! Leave it blank!)

Unit 1 — Number and Operations

Lesson 1: Number Properties

Lesson 2: Rational vs. Irrational Numbers

Lesson 3: Simplifying Square Roots

Lesson 4: Scientific Notation

Lesson 5: Fractions

Lesson 6: Order of Operations

Lesson 7: Percents

Lesson 8: Percent of Change

Lesson 9: Factoring

Lesson 10: Ratios and Proportions

Lesson 11: Direct and Inverse Variation

National Council of Teachers of Mathematics
Standards for School Mathematics

Number and Operations Standard for Grades 6–8
Expectations

Instructional programs from prekindergarten through grade 12 should enable all students to—	In grades 6–8 all students should—	The following lessons correspond to each expectation—
Understand numbers, ways of representing numbers, relationships among numbers, and number systems	• work flexibly with fractions, decimals, and percents to solve problems;	• Lessons 2, 5, 7, and 8
	• compare and order fractions, decimals, and percents efficiently and find their approximate locations on a number line;	• Lessons 2, 5, and 7
	• develop meaning for percents greater than 100 and less than 1;	• Lesson 7
	• understand and use ratios and proportions to represent quantitative relationships;	• Lesson 10
	• develop an understanding of large numbers and recognize and appropriately use exponential, scientific, and calculator notation;	• Lesson 4
	• use factors, multiples, prime factorization, and relatively prime numbers to solve problems;	• Lesson 9
	• develop meaning for integers and represent and compare quantities with them.	• Lesson 2
Understand meanings of operations and how they relate to one another	• understand the meaning and effects of arithmetic operations with fractions, decimals, and integers;	• Lessons 2, 5, and 6
	• use the associative and commutative properties of addition and multiplication and the distributive property of multiplication over addition to simplify computations with integers, fractions, and decimals;	• Lesson 1
	• understand and use the inverse relationships of addition and subtraction, multiplication and division, and squaring and finding square roots to simplify computations and solve problems.	• Lessons 3 and 6
Compute fluently and make reasonable estimates	• select appropriate methods and tools for computing with fractions and decimals from among mental computation, estimation, calculators or computers, and paper and pencil, depending on the situation, and apply the selected methods;	• Lessons 2 and 5
	• develop and analyze algorithms for computing with fractions, decimals, and integers and develop fluency in their use;	• Lessons 2 and 5
	• develop and use strategies to estimate the results of rational-number computations and judge the reasonableness of the results;	• Lesson 2
	• develop, analyze, and explain methods for solving problems involving proportions, such a scaling and finding equivalent ratios.	• Lesson 10

Lesson 1: Number Properties

▶ *Number properties* state a relationship between numbers.

1) The Commutative Property

The **commutative property** states that the *order* of the numbers when adding or multiplying does not change the sum or product:

addition	multiplication
$a + b = b + a$	$a \times b = b \times a$
$7 + 21 = 21 + 7$	$4 \times 2 = 2 \times 4$
$28 = 28$	$8 = 8$

Subtraction and division are *not* commutative. In other words, the order of the numbers when subtracting or dividing *does* change the answer:

subtraction	division
$a - b \neq b - a$	$a \div b \neq b \div a$
$11 - 4 \neq 4 - 11$	$36 \div 9 \neq 9 \div 36$
$7 \neq -7$	$4 \neq \dfrac{1}{4}$

2) The Associative Property

The **associative property** states that the *grouping* of numbers when adding or multiplying does not change the sum or the product:

addition	multiplication
$(a + b) + c = a + (b + c)$	$(a \times b) \times c = a \times (b \times c)$
$(2 + 3) + 4 = 2 + (3 + 4)$	$(4 \times 2) \times 5 = 4 \times (2 \times 5)$
$5 + 4 = 2 + 7$	$8 \times 5 = 4 \times 10$
$9 = 9$	$40 = 40$

Again, subtraction and division don't work with this property. Subtraction and division are *not* associative because the grouping of the numbers *does* change the answer:

subtraction	division
$(a - b) - c \neq a - (b - c)$	$(a \div b) \div c \neq a \div (b \div c)$
$(4 - 2) - 1 \neq 4 - (2 - 1)$	$(16 \div 4) \div 2 \neq 16 \div (4 \div 2)$
$2 - 1 \neq 4 - 1$	$4 \div 2 \neq 16 \div 2$
$1 \neq 3$	$2 \neq 8$

3) The Distributive Property

The **distributive property** relates the operations of multiplication and addition:

$$a(b + c) = ab + ac \qquad \text{or} \qquad (b + c)a = ab + ac$$

example 1

$3(2 + 6) = 3(2) + 3(6)$

$\quad 3(8) = 6 + 18$

$\quad\quad 24 = 24$

example 2

$4(x + 7) = 4(x) + 4(7)$

$\quad\quad\quad = 4x + 28$

Not understanding the distributive property can lead to a great number of problems when solving algebraic equations, so it is *very* important to know and understand this property!

example of common mistake

A common mistake is to distribute a number or variable but to forget to distribute the **sign** of that number or variable:

wrong way:	correct way:
$7 - 2(4 - 1) = 7 - 2(4) + 2(-1)$	$7 - 2(4 - 1) = 7 - 2(4) - 2(-1)$
$= 7 - 8 - 2$	$= 7 - 8 + 2$
$= -1 - 2$	$= -1 + 2$
$= -3$	$= 1$

▶ When an **identity** is combined with another number, that number stays the same.

4) Additive Identity Property

Zero is called the **additive identity** because adding *zero* to a number does not change it — *the number stays the same*:

$$a + 0 = a \qquad \text{or} \qquad 0 + a = a$$

(Note: Notice the *commutative property of addition* in the above equations. The order of the "0" and the "*a*" does not change the sum.)

5) Multiplicative Identity Property

One is called the **multiplicative identity** because multiplying any number by *one* does not change it — *the number stays the same*:

$$a \cdot 1 = a \qquad \text{or} \qquad 1 \cdot a = a$$

(Note: Notice the *commutative property of multiplication* in the above equations.)

▶ An **inverse** of something is its opposite, so the inverse of a mathematical operation undoes that operation: subtraction undoes addition, and division undoes multiplication.

6) Additive Inverse Property

When the **additive inverse** of a number is added to that number, *the sum will be <u>zero</u>*:

$$a + (-a) = 0 \qquad \text{or} \qquad (-a) + a = 0$$

So just as any number added to zero equals that number, adding the additive inverse of that number brings you back to zero.

example 3
$0 + 7 = 7$
$7 + (-7) = 0$

To find the additive inverse of a number, simply change the *sign* of that number.

7) Multiplicative Inverse Property

When a number is multiplied by its **multiplicative inverse**, *the product will be <u>one</u>*:

$$a \cdot \frac{1}{a} = 1 \qquad \text{or} \qquad \frac{1}{a} \cdot a = 1$$

(Note: $a \neq 0$)

The multiplicative inverse of a number is its **reciprocal** (where the **numerator** and **denominator** switch places).

---

example 4

What is the multiplicative inverse of 7?

The number 7 can be thought of as the fraction: $\frac{7}{1}$

If you switch the numerator and denominator, $\frac{7}{1}$ becomes $\frac{1}{7}$.

So the multiplicative inverse of 7 is $\frac{1}{7}$.

Checking the answer: If these two numbers are multiplied together, the product is one: $7 \times \frac{1}{7} = 1$.

Another way to think of a multiplicative inverse is how division is the opposite (or inverse) of multiplication. Just as 1 multiplied by any number equals that number, dividing by that number (or multiplying by the multiplicative inverse) brings you back to 1.

---

example 5

1) $1 \times 4 = 4$

$4 \div 4 = 1$

OR

2) $1 \times 4 = 4$

$4 \times \frac{1}{4} = 1$

▶ **Closure** refers to numbers within a *set*. (For example, the set of all positive numbers, set of all even numbers, set of all rational numbers, etc.)

8) Property of Closure

If a mathematical operation is performed on two numbers from a particular set, and the answer is also from that same set of numbers, then that set of numbers is *closed* under the operation performed.

---

example 6

If any two even numbers are added, the answer will always be an even number.

This means that the set of all even numbers is *closed under addition.*

---

example 7

Is the set of all even numbers closed under division?

Pick an even number and divide it by other even numbers to see if you can find a quotient that isn't even;

$8 \div 4 = 2$ in this case, the quotient is even, but

$8 \div 8 = 1$ and since 1 is an odd number, an example was found where dividing an even number by another even number resulted in an answer that was *not* even.

It just takes one occurrence to say that a set of numbers is *not* closed under a particular operation.

Therefore, the set of all even numbers is <u>not</u> *closed under division.*

▶ You saw that anything *added* to zero equals itself (the additive identity property), but what happens when something is *multiplied* by zero?

9) The Multiplicative Property of Zero

The **multiplicative property of zero** states that multiplying any number by zero equals zero:

$$a \cdot 0 = 0 \qquad \text{or} \qquad 0 \cdot a = 0$$

In multiplication, one number times another number can be thought of as having that first number a certain number of times (such as 4×3 means you have the number 4 three times: $4 + 4 + 4 = 12$).

If you have a number times zero, that means you have that number zero times.

▶ When two negative numbers are multiplied together, the answer is the same as multiplying the same positive numbers together.

10) $(-x)(-y) = xy$

To phrase it another way, *a negative times a negative equals a positive.*

This can be proven by using the additive inverse property, the multiplicative property of zero, and the distributive property:

1) From the additive inverse property: $b + (-b) = 0$

2) Multiply both sides by $-a$: $-a(b + (-b)) = -a \cdot (0)$

3) From the multiplicative property of zero: $-a \cdot (0) = 0$, so the equation in step 2 can be re-written as: $-a(b + (-b)) = 0$

4) From the distributive property: $-a(b + (-b)) = -a(b) + (-a)(-b)$, so the equation in step 3 can be re-written as: $-a(b) + (-a)(-b) = 0$, or as: $-ab + (-a)(-b) = 0$

5) Adding ab to both sides: $-ab + (-a)(-b) = 0$
$$\frac{+\ ab \qquad\qquad\qquad +\ ab}{0 + (-a)(-b) = ab}$$

6) Therefore, $(-a)(-b) = ab$, or if you replace a with x and replace b with y: $(-x)(-y) = xy$

This identity is mainly used when distributing and making sure you get the *signs* right, positive or negative.

example 8

Distribute: $(-2)(-x + 2)$

Using the distributive property, this can be re-written as: $(-2)(-x) + (-2)(2)$

Using the identity that was proven above, $(-2)(-x) = 2x$ and $(-2)(2) = -4$

Therefore, $(-2)(-x + 2) = 2x - 4$

Practice Problems

Name _____

Match each statement shown with the appropriate property:

1) $0 + d = d$

A) Commutative Property of Addition

2) $x + (y + z) = (x + y) + z$

B) Commutative Property of Multiplication

3) $6 \cdot \dfrac{1}{6} = 1$

C) Associative Property of Addition

4) $\dfrac{1}{2} + \left(-\dfrac{1}{2}\right) = 0$

D) Associative Property of Multiplication

5) $2 \cdot (3 \times 4) = (2 \times 3) \cdot 4$

E) Distributive Property

6) $6x + 3y = 3y + 6x$

F) Additive Identity Property

7) $7 \times 0 = 0$

G) Multiplicative Identity Property

8) $\dfrac{a}{3} \cdot \dfrac{3}{a} = 1$

H) Additive Inverse Property

9) $a(b + c) = a(c + b)$

I) Multiplicative Inverse Property

10) $8 \times 1 = 8$

J) Multiplicative Property of Zero

11) $(2xy)3 = 2x(y \cdot 3)$

12) $ab + cd = ba + cd$

13) $a(b + c) = ab + ac$

14) $4(xy) = 4(yx)$

15) $(2x + y) + 3 = 2x + (y + 3)$

16) $4 + 0 = 4$

17) $c + (-c) = 0$

18) $\dfrac{1}{b} \times 0 = 0$

19) $ab \times 1 = ab$

20) $-7(x - 1) = -7x + 7$

Simplify the following expressions using the distributive property:

21) $2(a + b)$

22) $-4(3m - 2n)$

23) $x - (y + 4)$

24) $-2(m + 2)$

25) $(a - 8)b$

26) $3x + 4y - 2(x - y)$

27) $-k(2k + 5)$

28) $a + 4(a + 2)$

Name the multiplicative inverse of each number or variable (assume that no variable represents zero):

29) 9

30) $\dfrac{1}{3}$

31) $\dfrac{2}{x}$

32) $5m$

33) $\dfrac{a + 2}{3}$

34) $2\dfrac{1}{4}$

Lesson 2: Rational vs. Irrational Numbers

Rational Numbers

A **rational number** is any number that can be expressed in the fractional form $\frac{a}{b}$, where a and b are both integers and the denominator is not equal to zero.

When a rational number is expressed as a decimal, the number will appear as a **terminating decimal**, which means the decimal stops, or as a **repeating decimal**, which means the decimal goes on forever, repeating the same values over and over.

terminating decimal

$$\frac{1}{4} = 0.25$$

$$\frac{2}{5} = 0.4$$

repeating decimal

$$\frac{1}{3} = 0.333.... \text{ or } 0.\overline{3}$$

$$\frac{4}{7} = 0.5714285714... \text{ or } 0.\overline{571428}$$

(While rational numbers can be expressed as decimals or fractions, this lesson will focus on decimals while Lesson 5 will explore fractions in more detail.)

Computing with Decimal Numbers

---

example 1

Compute:

$$3.62 + 49.1 + 56 =$$

When adding (or subtracting) rational numbers expressed as decimals, *always line up the decimal points vertically*, then add each column from right to left. If any numbers carry over (if a column adds up to more than 9) add the tens digit of that column to the next column (to the left). Make sure the decimal place in the answer lines up with the columns of numbers being added:

			1	
3.62	3.62	3.62	3.62	3.62
49.1 ⇒	49.1 ⇒	49.1 ⇒	49.1 ⇒	49.1
+ 56	+ 56	+ 56	+ 56	+ 56
	2	72	8.72	108.72

The answer is 108.72.

---

example 2

Compute:

$$43.68 \times 2.5 =$$

When multiplying decimal numbers, as opposed to when adding (or subtracting), do *not* line up the decimal points. Line up the numbers on the right:

$$
\begin{array}{r}
43.68 \\
\times \ \ 2.5 \\
\hline
\end{array}
$$

Next, multiply the bottom digit the farthest to the right by each digit in the top number, moving from right to left. If any product is greater than nine, write down the digit in the ones place and carry over the digit in the tens place. The carried-over digit will be added to the next product found:

$$
\begin{array}{r}
43.68 \\
\times \ \ 2.5 \\
\hline
\end{array}
\Rightarrow
\begin{array}{r}
{}^{4} \\
43.68 \\
\times \ \ 2.5 \\
\hline
0 \\
\end{array}
\Rightarrow
\begin{array}{r}
{}^{3} \\
43.68 \\
\times \ \ 2.5 \\
\hline
40 \\
\end{array}
\Rightarrow
\begin{array}{r}
{}^{1} \\
43.68 \\
\times \ \ 2.5 \\
\hline
840 \\
\end{array}
\Rightarrow
\begin{array}{r}
43.68 \\
\times \ \ 2.5 \\
\hline
21840 \\
\end{array}
$$

Before multiplying the next digit in the bottom number (the 2 in this case) by each digit in the top number, first place a zero in the first column (to the right), then multiply the same way as before:

$$
\begin{array}{r}
43.68 \\
\times \ \ 2.5 \\
\hline
21840 \\
0 \\
\end{array}
\Rightarrow
\begin{array}{r}
{}^{1} \\
43.68 \\
\times \ \ 2.5 \\
\hline
21840 \\
60 \\
\end{array}
\Rightarrow
\begin{array}{r}
{}^{1} \\
43.68 \\
\times \ \ 2.5 \\
\hline
21840 \\
360 \\
\end{array}
\Rightarrow
\begin{array}{r}
43.68 \\
\times \ \ 2.5 \\
\hline
21840 \\
7360 \\
\end{array}
\Rightarrow
\begin{array}{r}
43.68 \\
\times \ \ 2.5 \\
\hline
21840 \\
87360 \\
\end{array}
$$

Now add the two rows of numbers you just created and move the decimal point to the *left* the same number of places as the *total* number of digits that are to the *right* of the decimal in each of the numbers you just multiplied together. For example, 43.68 has two digits to the right of the decimal place, and 2.5 has one digit to the right of the decimal. So there are a total of three digits to the right of the decimal in those numbers:

$$
\begin{array}{r}
21840 \\
+ \ \ 87360 \\
\hline
109200 \\
\end{array}
$$

Moving the decimal point three places to the *left* gives us an answer of 109.2.

| example 3 |

Compute:

$$4.14 \div 1.8$$

When dividing with decimals (without a calculator), the number you are dividing by (called the **divisor**) needs to be a whole number. In this problem, the divisor is 1.8, which *not* a whole number. So, move the decimal point to the right until it *is* a whole number. [You can do this as long as you move the decimal point of the number you are dividing into (called the **dividend**) the same number of spaces to the right.]

$$1.8\overline{)4.14} \quad \Rightarrow \quad 18\overline{)41.4}$$

Now divide, placing the decimal point in the **quotient** (the answer) directly above the decimal point in the dividend:

$$
\begin{array}{c}
2. \\
18\overline{)41.4} \\
36
\end{array}
\Rightarrow
\begin{array}{c}
2. \\
18\overline{)41.4} \\
\underline{-36} \\
5
\end{array}
\Rightarrow
\begin{array}{c}
2. \\
18\overline{)41.4} \\
\underline{-36} \\
5\ 4
\end{array}
\Rightarrow
\begin{array}{c}
2.3 \\
18\overline{)41.4} \\
\underline{-36} \\
54 \\
54
\end{array}
\Rightarrow
\begin{array}{c}
2.3 \\
18\overline{)41.4} \\
\underline{-36} \\
54 \\
\underline{-54} \\
0
\end{array}
$$

The answer is 2.3.

| example 4 |

The difference between two temperature readings was 7 degrees. Which of the following could be the two temperature readings?

A. $-7°$ and $1°$
B. $-4°$ and $3°$
C. $-1°$ and $7°$
D. $-5°$ and $12°$

A **difference** refers to an answer found by *subtraction*. To see which answer has two numbers with a difference of 7, subtract each smaller number from the larger number in each answer.

What this question is really testing is to see if you understand how to subtract a negative number. When you subtract a negative number, it's the same as adding that same positive number:

$$a - (-b) = a + b$$

Applying this to each answer:

A) $1 - (-7) = 1 + 7 = 8$

B) $3 - (-4) = 3 + 4 = 7$

C) $7 - (-1) = 7 + 1 = 8$

D) $12 - (-5) = 12 + 5 = 17$

The answer is B.

Comparing Decimal Numbers

example 5

Erin determined the masses of some samples for her science project. The mass of each sample is listed below.

Sample	Mass (grams)
1	17
2	16.7
3	17.6
4	16.67

Which of the following correctly lists the samples in order from the **least** mass to the **greatest** mass?

A. 1, 2, 3, 4

B. 2, 3, 4, 1

C. 2, 4, 1, 3

D. 4, 2, 1, 3

Numbers can be ordered from smallest to largest by comparing the place value of each number. Going from left to right, check the tens places, ones place, tenths place, and so on, and see which digits are larger.

In this problem, all four numbers have a "1" in the tens place, so they can't be compared using the tens digit. Two of the numbers have a "7" in the ones place, while the other two numbers have a "6" in the ones place, so these first two numbers are bigger. Now we just need to figure out, among the two larger numbers (17 and 17.6), which one is larger than the other, and among the smaller numbers (16.7 and 16.67), which one is larger.

The next place value is the tenths place, so compare those digits. If there is no digit there, a "0" can be used for that place value. Since 6 is larger than 0, 17.6 is larger than 17 (or 17.0). And since 7 is larger than 6, 16.7 is larger than 16.67.

Listing all four numbers in order, from smallest to largest:

16.67, 16.7, 17, 17.6

The answer is D.

Irrational Numbers

An **irrational number** is any number that cannot be expressed as a rational number. There are two kinds of irrational numbers: irrational numbers known by other symbols, such as π (pi), or square roots that can't be simplified into rational numbers.

π is the ratio of the circumference of a circle to its diameter, and it equals 3.14159265358979323846.... (π is usually rounded off to 3.14.)

Pi is a number that, when expressed in decimal form, *never* ends, and *never* repeats. This is true for all irrational numbers, as any irrational number cannot be expressed as a terminating decimal or repeating decimal. (Note: Square roots are rational numbers if they are the square root of a rational number that's been squared.)

$$\sqrt{2} = 1.414213562...$$

$$\sqrt{3} = 1.732050808...$$

but

$$\sqrt{9} = 3$$

$$\sqrt{0.0625} = 0.25$$

---

example 6

Which of the following is an irrational number?

A. $\dfrac{4}{3}$

B. $\sqrt{24}$

C. $\sqrt{81}$

D. -4.07

A rational number can be expressed in the fractional form $\frac{a}{b}$ (where a and b are both integers and the denominator is not equal to zero). Any number expressed as a decimal is a rational number, which also includes integers. Square roots (and π) are typically the only irrational numbers you'll ever see.

Knowing this definition for a rational number, it's easy to eliminate answers A and D. The remaining answers are both square roots, but only one can be irrational if there is only one answer to this question.

The key to figuring out which square root is an irrational number is being able to recognize a **perfect square**, which is the square of an integer. (More on this in Lesson 3.) Since 81 is equal to 9 squared (or 9×9), $\sqrt{81}$ is equal to 9. Therefore, $\sqrt{81}$ is actually a rational number.

The answer is B.

Practice Problems

Name _____

Compute (*without* using a calculator):

1) $1.06 + 2.7 =$

2) $14 + 3.14 + 0.113 =$

3) $9.8 - 6.23 =$

4) $4.05 - 7.1 =$

5) $6.22 \times 3.7 =$

6) $20.7 \times 1.36 =$

7) $8.88 \div 2.4 =$

8) $11.25 \div 0.75 =$

Re-write the given numbers in order from smallest to largest:

9) 15.45, 14.54, 15.55, 14.94   10) 39, 38.9, 39.8, 38

11) 1,000,100; 1,000,001;   12) 14.89, 14.9, 14.889, 14.8
     1,001,000; 1,000,010

Label each number as *rational* or *irrational*:

13) 0   14) 3.14

15) $\sqrt{6}$ 16) $\dfrac{\sqrt{9}}{5}$

17) $\dfrac{1}{4}$ 18) π

19) $2.\overline{34}$ 20) $-\dfrac{2}{7}$

Lesson 3: Simplifying Square Roots

Understanding Square Roots

The **square root** of a given number is a value that, when multiplied by itself, produces that given number. Square roots come in pairs, a positive root and a negative root.

For example, a square root of 4 is 2, because $2 \times 2 = 4$. Another square root of 4 is -2, because $-2 \times -2 = 4$. This can be written as $\sqrt{4} = \pm 2$, which means that $\sqrt{4} = 2$ and $\sqrt{4} = -2$.

Note: You can *not* take the square root of a negative number.

To better understand square roots, it is very helpful to be familiar with perfect squares. A perfect square is the square of an integer:

perfect squares

$1^2 = 1$	$11^2 = 121$
$2^2 = 4$	$12^2 = 144$
$3^2 = 9$	$13^2 = 169$
$4^2 = 16$	$14^2 = 196$
$5^2 = 25$	$15^2 = 225$
$6^2 = 36$	$16^2 = 256$
$7^2 = 49$	$17^2 = 289$
$8^2 = 64$	$18^2 = 324$
$9^2 = 81$	$19^2 = 361$
$10^2 = 100$	$20^2 = 400$

example 1

Which of the following statements about $\sqrt{121}$ is **not** true?

A. $\sqrt{121}$ is an irrational number.

B. $\sqrt{121}$ is an integer.

C. $\sqrt{121}$ is a real number.

D. $\sqrt{121}$ is a rational number.

To answer this question, knowledge of perfect squares is extremely helpful. A square root is an irrational number, *unless it is the square root of a perfect square*:

$$11^2 = 121 \qquad so \qquad \sqrt{121} = 11$$

11 is an *integer* and a *real number* and a *rational number*, but 11 is **not** an *irrational number*, so $\sqrt{121}$ is not an irrational number.

The answer is A.

<u>Simplifying Square Roots</u>

To simplify the square root of a number, find two factors of that number, *one of which is a perfect square.*

(A **factor** is an integer that divides evenly into another number.)

example 2

Simplify $\sqrt{18}$.

$$\sqrt{18} = \sqrt{9 \times 2}$$
$$= \sqrt{9} \times \sqrt{2}$$
$$= 3 \times \sqrt{2}$$
$$= 3\sqrt{2}$$

This example demonstrates the following property of square roots:

$$\sqrt{ab} = \sqrt{a} \cdot \sqrt{b}$$

This applies to fractions as well:

$$\sqrt{\frac{a}{b}} = \frac{\sqrt{a}}{\sqrt{b}}$$

example 3

Simplify $\sqrt{\dfrac{4}{9}}$.

$$\sqrt{\frac{4}{9}} = \frac{\sqrt{4}}{\sqrt{9}} = \frac{2}{3}$$

example 4

Simplify $\sqrt{\dfrac{32}{20}}$.

$$\sqrt{\frac{32}{20}} = \frac{\sqrt{32}}{\sqrt{20}} = \frac{\sqrt{16 \cdot 2}}{\sqrt{4 \cdot 5}} = \frac{\sqrt{16} \cdot \sqrt{2}}{\sqrt{4} \cdot \sqrt{5}} = \frac{4\sqrt{2}}{2\sqrt{5}} = \frac{2\sqrt{2}}{\sqrt{5}}$$

It's not acceptable to leave a square root (also called a **radical**) in the denominator of a fraction. The way to leave an answer without a radical in the denominator is to multiply the numerator and denominator by the same radical that is in the denominator.

$$\frac{2\sqrt{2}}{\sqrt{5}} \cdot \frac{\sqrt{5}}{\sqrt{5}} = \frac{2\sqrt{2} \cdot \sqrt{5}}{\sqrt{5} \cdot \sqrt{5}} = \frac{2\sqrt{2 \cdot 5}}{\sqrt{5 \cdot 5}} = \frac{2\sqrt{10}}{\sqrt{25}} = \frac{2\sqrt{10}}{5}$$

This is called *rationalizing the denominator.*

Estimating the Value of a Square Root

Perfect squares can also be used to estimate the approximate value of a square root.

---

example 5

Which is the **best** approximation of $\sqrt{72}$?

A. 7.2

B. 9.1

C. 8.9

D. 8.5

To answer this question, make a list of perfect squares close to 72.

$7^2 = 49$		$\sqrt{49} = 7$
$8^2 = 64$	or	$\sqrt{64} = 8$
$9^2 = 81$		$\sqrt{81} = 9$

72 is between 64 and 82, so $\sqrt{72}$ is between $\sqrt{64}$ and $\sqrt{81}$, which means $\sqrt{72}$ is between 8 and 9. That eliminates A and B as possible answers. And 72 is almost exactly between 64 and 81, so $\sqrt{72}$ is closer 8.5 than 8.9.

The answer is D.

---

example 6

The square root of 31 is between which two whole numbers?

A. 4 and 5

B. 5 and 6

C. 6 and 7

D. 7 and 8

This problem can also be solved by first making a list of perfect squares.

$$4^2 = 16 \qquad\qquad \sqrt{16} = 4$$
$$5^2 = 25 \qquad\qquad \sqrt{25} = 5$$
$$6^2 = 36 \qquad\text{or}\qquad \sqrt{36} = 6$$
$$7^2 = 49 \qquad\qquad \sqrt{49} = 7$$
$$8^2 = 64 \qquad\qquad \sqrt{64} = 8$$

31 is between 25 and 36, so the square root of 31 is between $\sqrt{25}$ and $\sqrt{36}$, or between 5 and 6.

The answer is B.

example 7

A part of the real number line is shown below.

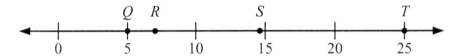

Which letter best represents the location of $\sqrt{50}$?

A. Q
B. R
C. S
D. T

Once again, a list of perfect squares will help, this time near 50.

$$6^2 = 36 \qquad\qquad \sqrt{36} = 6$$
$$7^2 = 49 \qquad\text{or}\qquad \sqrt{49} = 7$$
$$8^2 = 64 \qquad\qquad \sqrt{64} = 8$$

50 is really close to 49, so $\sqrt{50}$ will be close to $\sqrt{49}$, or 7. On the number line above, only one point looks like it could represent a value near 7.

The answer is B.

Practice Problems

Name _____

Simplify. Leave in radical form (in other words, don't use a calculator).

1) $\sqrt{20}$

2) $-\sqrt{64}$

3) $\sqrt{900}$

4) $\sqrt{0.16}$

5) $-\sqrt{0.0025}$

6) $\sqrt{\dfrac{16}{25}}$

7) $\sqrt{\dfrac{3}{27}}$

8) $\sqrt{\dfrac{4}{3}}$

9) $\dfrac{3}{\sqrt{7}}$

10) $\sqrt{8}\cdot\sqrt{32}$

11) $5\sqrt{14}\cdot2\sqrt{7}$

12) $-3\sqrt{5}\cdot\sqrt{\dfrac{4}{5}}$

Name the integers between which each value lies (without using a calculator).

13) $\sqrt{11}$

14) $\sqrt{34}$

15) $2\sqrt{3}$

16) $3\sqrt{6}$

17) $\dfrac{\sqrt{25}}{2}$

18) $\dfrac{\sqrt{14}}{2}$

Label each statement as *true* or *false* (without using a calculator).

19) $2 < \sqrt{7} < 3$

20) $6^2 < 65 < 7^2$

21) $2^2 < \sqrt{20} < 3^2$

22) $\sqrt{45} < 10 < \sqrt{90}$

Lesson 4: Scientific Notation

Very large or very small numbers can be difficult to write out. Scientific notation is a way of expressing numbers without writing out too many digits.

A number written in scientific notation is expressed in the form:

$$a \times 10^n \text{ where } 1 \leq a < 10 \text{ and } n \text{ is an integer}$$

example 1

Write 1,700,000 in scientific notation.

1) Move the decimal point to the left until you have a number between 1 and 10.

  $1,700,000 \rightarrow 1.7$

2) This number is a: $a = 1.7$

3) The number of places you moved the decimal point to the *left* is n.

  $1,700,000 = 1.7 \times 10^6$

example 2

Write 0.000078 in scientific notation.

1) Move the decimal point to the right until you have a number between 1 and 10.

  $0.000078 \rightarrow 7.8$

2) This number is a: $a = 7.8$

3) The number of places you moved the decimal point to the *right* is $-n$.

  $0.000078 = 7.8 \times 10^{-5}$

Products and quotients of numbers expressed in scientific notation can be found by following the rules of exponents for multiplying and dividing monomials (which will be explained in detail in Lesson 24):

$$a^m \cdot a^n = a^{m+n}$$

$$\frac{a^m}{a^n} = a^{m-n}$$

example 3

$$a^2 \cdot a^4 = a^{2+4} = a^6$$

You can check this answer by writing out all of the variables:

$$a^2 = a \cdot a$$
$$a^4 = a \cdot a \cdot a \cdot a$$
$$(a \cdot a) \times (a \cdot a \cdot a \cdot a) = a \cdot a \cdot a \cdot a \cdot a \cdot a \quad \text{or} \quad a^6$$

This works in scientific notation as well.

example 4

$$10^2 \cdot 10^4 = 10^6$$

This can be written out the same way:

$$10^2 = 10 \cdot 10$$
$$10^4 = 10 \cdot 10 \cdot 10 \cdot 10$$
$$(10 \cdot 10) \times (10 \cdot 10 \cdot 10 \cdot 10) = 10 \cdot 10 \cdot 10 \cdot 10 \cdot 10 \cdot 10 \quad \text{or} \quad 10^6$$

example 5

$$\frac{a^5}{a^2} = a^{5-2} = a^3$$

Writing out all of the variables:

$$\frac{a^5}{a^2} = \frac{a \cdot a \cdot a \cdot a \cdot a}{a \cdot a}$$

The two variables in the denominator cancel out two of the variables in the numerator:

$$\frac{a \cdot a \cdot a \cdot a \cdot a}{a \cdot a} = \frac{a \cdot a \cdot a}{1} = a^3$$

Again, this also works in scientific notation:

example 6

$$\frac{10^5}{10^2} = 10^{5-2} = 10^3$$

26

The rules of exponents (for how they apply to scientific notation) can then be used in other problems:

example 7

Divide: $\dfrac{5.4 \times 10^6}{2.7 \times 10^4}$

$$\frac{5.4 \times 10^6}{2.7 \times 10^4} = \frac{5.4}{2.7} \times \frac{10^6}{10^4}$$

$$= 2 \times 10^2$$

$$= 200$$

example 8

Multiply: $(1300)(0.000006)$

$$(1300)(0.000006) = (1.3 \times 10^3)(6 \times 10^{-6})$$

$$= (1.3 \times 6)(10^3 \times 10^{-6})$$

$$= 7.8 \times 10^{-3}$$

example 9

Which of the following measurements would **most likely** be given with a **negative** exponent in scientific notation?

A. the diameter of a blood cell in centimeters
B. the distance to the Sun in kilometers
C. the weight of a pencil in grams
D. the mass of a rocket in kilograms

A negative exponent in scientific notation means that, in standard notation, the number is smaller than 1. A positive exponent would mean a value larger than 1.

$$\text{(Remember: } 10 = 1 \times 10^1,\ 1 = 1 \times 10^0,\ 0.1 = 1 \times 10^{-1}\text{)}$$

This question also requires you to have an understanding of several metric measurements, such as what a centimeter, kilometer, gram, and kilogram are.

The important thing here is which is bigger, the object or the unit of measurement. If the object is bigger, it will take more than one of the unit of measurement to measure it. If the unit of measurement is bigger than the object, it will take a fraction of the unit of measurement to describe it.

Investigating each answer:

Answer A: A blood cell is very tiny and invisible to the eye without a microscope. A centimeter is larger than that, so a fraction of a centimeter would be needed to measure one cell.

Answer B: A kilometer is a little more than half a mile (1 kilometer = 0.62 miles). The distance to the sun is much farther than that; it would take many, many kilometers to describe this distance.

Answer C: A pencil doesn't weigh very much, but how does it compare to a gram? A *kilo*gram is just over 2 pounds (1 kilogram = 2.2 pounds), so a gram is *one-thousandth* of that. Since Answer A already seems like the best answer, if you guess that a pencil weighs more than one gram, that would be a good guess.

Answer D: It's already been mentioned that a kilogram is a little more than 2 pounds, and a rocket definitely weighs much more than that. This means it would take many kilograms to measure the mass of a rocket.

The answer is A.

---

example 10

Which is closest to $2 \cdot 9^5$?

A. 1,000
B. 10,000
C. 100,000
D. 1,000,000

In this problem, by asking which answer is "closest" to $2 \cdot 9^5$, you are being asked to make an approximation, so round the number 9 to 10.

$$2 \cdot 9^5 \text{ then becomes } 2 \cdot 10^5$$

$2 \cdot 10^5 = 200,000$, so $2 \cdot 9^5$ must be closer to 100,000 than 10,000 or 1,000,000.

The answer is C.

Practice Problems

Name _____

Write each number in scientific notation.

1) 4500

2) 77,100

3) 6,000,000,000

4) 0.0059

5) 0.000482

6) 0.02300

7) 304×10^{-5}

8) $19,300 \times 10^{2}$

9) 0.0043×10^{-3}

Write each number in standard notation.

10) 3.740×10^{7}

11) 4.66×10^{2}

12) 8×10^{6}

13) 6.4×10^{-4}

14) 9.45×10^{-2}

15) 2.5×10^{-6}

16) 19×10^{4}

17) 256×10^{3}

18) 47.3×10^{-2}

Evaluate each expression and write the answer in scientific *and* standard notation.

19) (4500)(100)

20) (54,000)(710)

21) (0.003)(0.0018)

22) (0.075)(0.00005)

23) $\dfrac{0.00216}{36}$

24) $\dfrac{1.8 \times 10^5}{0.9 \times 10^2}$

25) $\dfrac{7.2 \times 10^{-3}}{24 \times 10^{-7}}$

26) $\dfrac{4200}{2.1 \times 10^{-3}}$

Lesson 5: Fractions

To be successful in understanding algebraic concepts in general, a thorough understanding of fractions is necessary.

<u>Adding and Subtracting Fractions</u>

To add or subtract fractions, a **common denominator** is required, and then the numerators can be added or subtracted while the denominators remain the same.

example 1

Add: $\dfrac{2}{5} + \dfrac{1}{5}$

Because the denominators are already the same, the numerators can be added:

$$\frac{2}{5} + \frac{1}{5} = \frac{2+1}{5} = \frac{3}{5}$$

example 2

Add: $\dfrac{2}{5} + \dfrac{1}{3}$

In this problem, a common denominator is needed before adding the fractions. The easiest way to find a common denominator is to multiply the two denominators together:

$$5 \times 3 = 15 \leftarrow \text{common denominator}$$

To change the denominator of each fraction without changing the value of the *entire* fraction, *multiply the numerator **and** denominator by the same number*:

$$\frac{2}{5} \times \left(\frac{3}{3}\right) + \frac{1}{3} \times \left(\frac{5}{5}\right) = \frac{2\times3}{5\times3} + \frac{1\times5}{3\times5} = \frac{6}{15} + \frac{5}{15} = \frac{6+5}{15} = \frac{11}{15}$$

example 3

Add: $\dfrac{5}{12} + \dfrac{1}{6}$

Multiplying 12 and 6 would result in a common denominator, but the **lowest common denominator** is easier to use.

If the two denominators have any **common factors**, multiply each denominator by the *uncommon* factors of the other denominator.

$$\frac{5}{12} + \frac{1}{6}$$

$$\frac{5}{2\times 6} + \frac{1}{6}$$

$$\frac{5}{2\times 6} + \frac{1}{6} \times \left(\frac{2}{2}\right)$$

$$\frac{5}{12} + \frac{2}{12} = \frac{5+2}{12} = \frac{7}{12}$$

example 4

Add: $\dfrac{3}{10} + \dfrac{7}{15}$

$$\frac{3}{5\times 2} + \frac{7}{5\times 3}$$

2 and 3 are the uncommon factors of the two denominators.

$$\frac{3}{5\times 2} \times \left(\frac{3}{3}\right) + \frac{7}{5\times 3} \times \left(\frac{2}{2}\right)$$

$$\frac{9}{30} + \frac{14}{30} = \frac{9+14}{30} = \frac{23}{30}$$

Subtracting fractions works the same exact way — a common denominator is required — and then the numerators (and only the numerators) can be subtracted.

Multiplying and Dividing Fractions

To multiply two fractions, multiply straight across: numerator times numerator, and denominator times denominator.

$$\frac{a}{b} \times \frac{c}{d} = \frac{ac}{bd}$$

example 5

Multiply: $\dfrac{2}{3} \times \dfrac{4}{7}$

$$\frac{2}{3} \times \frac{4}{7} = \frac{2 \times 4}{3 \times 7} = \frac{8}{21}$$

To divide fractions, take the reciprocal of the *second* fraction, then multiply the fractions.

$$\frac{a}{b} \div \frac{c}{d} = \frac{a}{b} \times \frac{d}{c} = \frac{ad}{bc}$$

example 6

Divide: $\dfrac{6}{15} \div \dfrac{2}{3}$

$$\frac{6}{15} \div \frac{2}{3} = \frac{6}{15} \times \frac{3}{2} = \frac{6 \times 3}{15 \times 2} = \frac{18}{30}$$

Reducing Fractions

To reduce a fraction:

1) Find the **greatest common factor** (GCF) of the numerator and the denominator.

2) Divide the numerator and denominator by that GCF.

example 7

Simplify: $\dfrac{18}{30}$

The GCF of 18 and 30 is 6:

$$18 = 6 \times 3$$

$$30 = 6 \times 5$$

Divide the numerator and denominator by the GCF, which is 6 in this problem:

$$\frac{18 \div 6}{30 \div 6} = \frac{3}{5}$$

$$\frac{18}{30} = \frac{3}{5}$$

<u>Comparing Fractions</u>

There are **three** ways to figure out how to order fractions from smallest to largest:

1) Re-write all of the fractions with a common denominator. Once this is done, the numerators can be compared.

---

example 8

Arrange the following fractions from smallest to largest.

$$\frac{5}{8}, \frac{3}{4}, \frac{2}{3}$$

24 is the **least common multiple** of 8, 4, 3:

$$8, 16, \textbf{24}, \dots$$

$$4, 8, 12, 16, 20, \textbf{24}, \dots$$

$$3, 6, 9, 12, 15, 18, 21, \textbf{24}, \dots$$

Multiply each fraction so that each has a denominator of 24:

$$\frac{5}{8} \times \left(\frac{3}{3}\right) = \frac{15}{24}, \quad \frac{3}{4} \times \left(\frac{6}{6}\right) = \frac{18}{24}, \quad \frac{2}{3} \times \left(\frac{8}{8}\right) = \frac{16}{24}$$

Comparing these fractions gives us:

$$\frac{15}{24}, \frac{16}{24}, \frac{18}{24}$$

These can then be reduced back to:

$$\frac{5}{8}, \frac{2}{3}, \frac{3}{4}$$

---

2) Compare the fractions two at a time by comparing their **cross products**.

---

example 9

Arrange the following fractions from smallest to largest.

$$\frac{5}{8}, \frac{3}{4}, \frac{2}{3}$$

First compare $\dfrac{5}{8}$ and $\dfrac{3}{4}$:

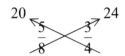

since $20 < 24$, then $\dfrac{5}{8} < \dfrac{3}{4}$

Now compare $\dfrac{3}{4}$ and $\dfrac{2}{3}$:

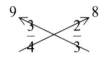

since $9 > 8$, then $\dfrac{3}{4} > \dfrac{2}{3}$

Next compare $\dfrac{5}{8}$ and $\dfrac{2}{3}$:

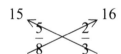

since $15 < 16$, then $\dfrac{5}{8} < \dfrac{2}{3}$

After these comparisons, the fractions can be written in order as:

$$\frac{5}{8}, \frac{2}{3}, \frac{3}{4}$$

3) Convert the fractions to decimals and compare the values. This is the fastest method for comparing fractions, especially if you have a calculator.

---

example 10

Arrange the following fractions from smallest to largest.

$$\frac{5}{8}, \frac{3}{4}, \frac{2}{3}$$

$$\frac{5}{8} = 0.625$$

$$\frac{2}{3} = 0.666\ldots$$

$$\frac{3}{4} = 0.75$$

If you don't have a calculator, then convert the fractions to decimals by writing out the long division:

```
      0.625           0.666          0.75
  8 ) 5.000       3 ) 2.000      4 ) 3.00
     -48             -18            -28
      20              20             20
     -16             -18            -20
      40              20              0
     -40             -18
       0               2
```

Practice Problems

Name _____

Reduce the following fractions:

1) $\dfrac{15}{27}$

2) $\dfrac{18}{48}$

3) $\dfrac{34}{85}$

Add the fractions and then reduce to lowest terms:

4) $\dfrac{1}{4} + \dfrac{3}{8} =$

5) $\dfrac{1}{6} + \dfrac{5}{8} =$

6) $\dfrac{2}{5} + \dfrac{3}{7} =$

Subtract and then reduce to lowest terms:

7) $\dfrac{5}{6} - \dfrac{2}{3} =$

8) $\dfrac{1}{2} - \dfrac{1}{3} =$

9) $\dfrac{9}{10} - \dfrac{3}{4} =$

Multiply and then reduce to lowest terms:

10) $\dfrac{2}{3} \times \dfrac{1}{6} =$

11) $\dfrac{7}{9} \times \dfrac{3}{5} =$

12) $\dfrac{6}{11} \times \dfrac{3}{4} =$

Divide and then reduce to lowest terms:

13) $\dfrac{1}{2} \div \dfrac{1}{5} =$

14) $\dfrac{8}{9} \div \dfrac{2}{3} =$

15) $\dfrac{3}{4} \div \dfrac{5}{8} =$

Replace each ? with <, >, or = to make each sentence true:

16) $\dfrac{4}{9} \; \underline{?} \; \dfrac{5}{11}$

17) $\dfrac{3}{8} \; \underline{?} \; \dfrac{9}{24}$

18) $\dfrac{8}{11} \; \underline{?} \; \dfrac{12}{17}$

Rewrite the fractions in order from least to greatest (without using a calculator):

19) $\dfrac{2}{5}, \dfrac{4}{9}, \dfrac{3}{7}$

20) $\dfrac{1}{4}, \dfrac{3}{13}, \dfrac{2}{9}$

21) $\dfrac{5}{8}, \dfrac{4}{7}, \dfrac{3}{5}$

Lesson 6: Order of Operations

When simplifying numerical expressions, you must follow rules called the **order of operations**. They work as follows:

1) Perform operations within grouping symbols — parentheses ( ), brackets [ ], braces { }, radicals $\sqrt{}$, absolute value bars | |, or in the numerator or denominator of a fraction.

2) Simplify all exponents (including square roots, cube roots, or other radicals).

3) Multiply and/or divide terms in order from left to right.

4) Add and/or subtract terms in order from left to right.

An east way of remembering the order of operations for simplifying numerical expressions is with the abbreviation PEMDAS:

P — parentheses
E — exponents
M — multiplication
D — division
A — addition
S — subtraction

Parentheses

example 1

Compute:

$$8 - (-5 + 3 \times 7) =$$

The first expression to simplify is the one in parentheses:

$$(-5 + 3 \times 7)$$

The rules for order of operations applies to this expression, where any multiplication must happen first (since there are no exponents):

$$(-5 + 21)$$

Now these numbers can be added:

$$(16)$$

The remainder of the original expression can be simplified now:

$$8 - (16) = -8$$

The answer is -8.

Absolute Value

example 2

What is the value of the following expression?

$$\left|-5\right| + \left|-5\right| - \left|-3\right|$$

The **absolute value** of a number is its distance from zero, which can be seen visually on a number line:

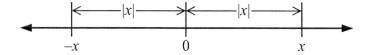

What that means mathematically is, the positive and negative forms of any number have the same absolute value:

$$\left|a\right| = \left|-a\right| = a$$

From the above expression, $\left|-5\right| = 5$, and $\left|-3\right| = 3$, so the expression can be rewritten as:

$$5 + 5 - 3$$

Notice that only the *signs* of each number changed, but the subtraction of the third term is still there. Only minus signs *inside* absolute value symbols are changed.

The expression can then now be simplified:

$$5 + 5 - 3 = 10 - 3 = 7$$

The answer is 7.

Exponents

---

example 3

Compute:

$$(-4)^3 =$$

An exponent indicates how many of something is being multiplied together. So -4 raised to the third power is the same as:

$$(-4) \times (-4) \times (-4)$$

The only operation being applied here is multiplication, so you can multiply from left to right (and remember that a negative times a negative equals a positive):

$$(-4) \times (-4) \times (-4) = 16 \times (-4) = \mathbf{-64}$$

The answer is -64.

(Note: When a negative number is raised to a odd power, or in other words, when the exponent is an odd number, the result will be negative. But when a negative number is raised to an even power, the result will be positive.)

---

example 4

What is the value of the expression $3(2-4)^2 + 3$?

A. -33
B. -9
C. 15
D. 39

In the problem, just go through the order of operations (PEMDAS).

Parentheses first: $3(2-4)^2 + 3 = 3(-2)^2 + 3$

Exponents next: $3(-2)^2 + 3 = 3(4) + 3$

then any Multiplication or Division: $3(4) + 3 = 12 + 3$

and any Addition or Subtraction last: $12 + 3 = 15$

The answer is C.

Evaluating Algebraic Expressions

example 5

If $x = 2$ and $y = 4$, what is the value of the following expression?

$$x - 5y$$

A. -18
B. -12
C. 12
D. 18

This problem requires you to substitute numbers into each variable and then simplify the expression. Without any numbers to substitute, this expression is as simplified as it can be, but when we put 2 in for x, and 4 in for y, the numerical expression we get *can* be simplified, using the rules for the order of operations of course:

$$(2) - 5(4) = 2 - 20 = -18$$

The answer is A.

Practice Problems

Name _____

Simplify each expression:

1) $(6 + 2)^2$

2) $3^3 - 3^2 \times 3$

3) $8 \div 2 \times 3 \div 6$

4) $|8 - 5| + |5 - 8|$

5) $8 \cdot 2^2 + 5 - (2^3 - 4)$

6) $15 \div 3 + 4(3 - 2)^2$

7) $(2 + (2^3 \div 2^2))^2$

8) $30 \div (4 + 2(7 - 4)^2 - 3 \cdot 4)$

9) $4 \cdot (6 - 2) + (3 + 1)^2$

10) $\dfrac{\left|4^2 - 6^2\right|}{(8 - 2) \div 2} - 2^3$

Simplify each expression:

11) $3 + 4 \cdot 5 - 6 \cdot 7 \div 3$

12) $\sqrt{(5-2)^2 + (8-6)^2}$

13) $\dfrac{\sqrt{3 + (2 \times 11)}}{(11 + 4) \div 3}$

14) $2 + [3 + (27 \div 9)]^2$

15) $5^2 \div 5 + 2^2 \cdot 6 - 20 \div 5$

16) $\dfrac{2^3 - 3^2}{4 - 5} + (6 - 4) \div 2$

Simplify each expression if $a = 1$, $b = 2$, $c = 3$, $x = 4$, $y = 5$, and $z = 6$:

17) $|b - z| + |a - y|$

18) $(a - x)^2 \div (b + 1)$

19) $\dfrac{(x - a)^2}{c} + \dfrac{(b - z)^2}{x}$

20) $\sqrt{(y + 2) - (y - 2)}$

Lesson 7: Percents

A percent is derived from a ratio of one number to another. When that ratio is set equal to another ratio of some **rate** to 100, this rate is also called a **percent**.

$$\frac{4}{5} = 0.80 \qquad \text{or} \qquad \frac{4}{5} = \frac{80}{100}$$

$$\Rightarrow \frac{4}{5} = 80\%$$

(*Per cent* means "per hundred", that's why there are 100 *cents* in one dollar.)

To change a decimal to a percent, the decimal point is moved *two* places to the **right**.

---
example 1

Express 0.56 as a percent.

$$0.56 = 56\%$$

To convert a percent back into a decimal, move the decimal point *two* places to the **left**.

---
example 2

Express 29% as a decimal.

$$29\% = 0.29$$

To change a ratio (or fraction) into a percent, first divide the denominator into the numerator to get a decimal value, and then move the decimal point.

---
example 3

Express $\frac{1}{5}$ as a percent.

$$\frac{1}{5} = 0.20 \qquad \text{or} \qquad 5\overline{)1.0}^{\,0.2}$$

$$0.20 = 20\%$$

So $\frac{1}{5} = 20\%$.

Any number expressed as a percent can also be converted to a fraction.

---

example 4

Express 75% as a fraction.

First, convert the percent to a decimal by moving the decimal point two places to the left.

$$75\% = 0.75$$

Next, rewrite the decimal as a fraction.

$$0.75 = \frac{75}{100}$$

Last, reduce the fraction as much as possible.

$$\frac{75}{100} = \frac{25 \times 3}{25 \times 4} = \frac{3}{4}$$

$$\text{So } 75\% = \frac{3}{4}$$

Solving Problems that Involve Percent

When setting up an equation, there are certain key words to look for.

"is" means "equals"

"of" means "times" (multiplication)

"what" means the unknown variable

---

example 5

What is 30% of 600?

To answer this question, convert it into an equation to be solved.

what = "x"	is = "="	30% = "0.30"
of = "times" or "×"	600 = "600"	

Converting the sentence into an equation:

$$x = 0.30 \times 600$$

$$x = 180$$

180 is 30% of 600.

Practice Problems

Name _____

Convert the following decimals to percents:

1) 0.43

2) 0.76

3) 1.01

4) 0.002

5) 0.4

6) 3.11

7) 5.7

8) 1.234

9) 4

Convert the following percents to decimals:

10) 75%

11) 9%

12) 100%

13) 50%

14) 6.5%

15) 33.3%

16) 1001%

17) $\frac{1}{2}$%

18) 46%

Convert the following fractions to percents:

19) $\frac{2}{5}$

20) $\frac{5}{4}$

21) $\frac{7}{20}$

22) $\frac{1}{6}$

23) $\frac{5}{8}$

24) $\frac{3}{4}$

25) $\frac{4}{9}$

26) $\frac{7}{10}$

27) $\frac{8}{13}$

Convert the following percents to fractions (in reduced form):

28) 16%

29) 50%

30) 150%

31) 33.$\overline{3}$%

32) 40%

33) 4%

34) 12.5%

35) 5%

36) 75%

Convert each sentence to an equation and then solve:

37) What is 20% of 25?

38) What is 40% of 40?

39) 15 is what percent of 60?

40) 30% of what number is 12?

41) What percent of 90 is 45?

42) What percent of 10 is 12?

48

Lesson 8: Percent of Change

When the value of something increases or decreases, that change can be expressed as a percent. This **percent of change** can be calculated using the following formula:

$$\text{percent of change} = \frac{\text{new number} - \text{original number}}{\text{original number}}$$

By always writing the "new number" first in the numerator, the *sign* of the resulting answer will indicate if it is a **percent increase** (positive sign) or **percent decrease** (negative sign).

---

example 1

A CD at a local store had its price changed from $10 to $8. What was the percent of change of the price of the CD?

the "original number" is 10

the "new number" is 8

Substitute these into the formula for percent of change:

$$\text{percent of change} = \frac{8 - 10}{10} = \frac{-2}{10} = -0.20 = -20\%$$

There was a 20% decrease in the price of the CD.

---

example 2

After stopping at a bus stop, the number of students sitting on the bus changed from 16 to 20. What is the percent of change in the number of students on the bus?

the "original number" is 16

the "new number" is 20

Substitute these into the formula for percent of change:

$$\text{percent of change} = \frac{20 - 16}{16} = \frac{4}{16} = 0.25 = 25\%$$

There was a 25% increase in the number of students on the bus.

Note: As you may have noticed, the percent of change depends on what the *original number* was since that is the value that goes into the denominator.

---

example 3

The percent of change from 12 to 16 is different than the percent of change from 16 to 12.

$$12 \rightarrow 16 \qquad\qquad 16 \rightarrow 12$$

original number = 12   original number = 16
new number = 16   new number = 12

$$\text{percent of change} = \frac{16-12}{12} \qquad \text{percent of change} = \frac{12-16}{16}$$

$$= \frac{4}{12} \qquad\qquad = \frac{-4}{16}$$

$$= 0.333\ldots \qquad\qquad = -0.25$$

$$= 33\% \text{ increase} \qquad\qquad = 25\% \text{ decrease}$$

Another kind of problem involving percent of change is when the percent of increase or decrease is provided and you have to determine the value of the original number or new number.

---

example 4

The regular price of a CD player is $74. It is on sale for 20% off. Which of the following is **closest** to the sale price?

A. $40

B. $50

C. $60

D. $70

A percentage *of* a number means that multiplication is involved (remember that "of" means "times".)

So 20% off is 20% *of* $74, and that value is then subtracted from $74:

20% of 74 $\Rightarrow$ 0.20 × 74 = 14.80

74 − 14.80 = 59.20

The sale price is close to $60

The answer is C.

Other important key words to notice when setting up a problem involving percent of change are "less than" or "more than":

"Less than" refers to *subtraction* and "more than" refers to *addition*.

Six *less than* ten is four. $\Rightarrow 10 - 6 = 4$
Seven *more than* five is twelve. $\Rightarrow 5 + 7 = 12$

"IS less than" is represented by "<" and "IS more than" is represented by ">".

Two *is less than* six. $\Rightarrow 2 < 6$
Eleven *is more than* x. $\Rightarrow 11 > x$

Don't get these confused, and remember that a percent in a math problem is a percent *of* something.

---

example 5

Thirty is twenty percent more than what number?

Rewrite this as follows:

30 = some number + 20% of that number

$30 = x + 0.20x$

$30 = 1.2x$

$\dfrac{30}{1.2} = \dfrac{1.2x}{1.2}$

$x = 25$

30 is 20% more than 25.

This result can be confirmed using the formula for percent of change.

the "original number" is 25, and the "new number" is 30

Substitute these into the formula for percent of change:

$$\text{percent of change} = \frac{30 - 25}{25} = \frac{5}{25} = 0.20 = 20\%$$

Percent of Change Over Time

---

example 6

An automobile is purchased for $18,000. Its value decreases each year according to the following schedule:

• The car's value decreases by 30% in the first year.

• After the first year, its value decreases by 20% each year.

---

a. What is the value of this car at the end of one year? Explain or show how you found your answer.

After one year, the car has a value that is 30% less than its original value. To find the number 30% less than $18,000, multiply $18,000 by 30% (or 0.30) and then subtract that value from $18,000:

$$\$18,000 - (0.30)(\$18,000) = \$18,000 - \$5400$$

$$= \$12,600$$

b. During which year will the car's value decrease to less than half its original price? Explain or show how you found your answer.

To calculate the value of the car in future years, 20% of the car's value is subtracted each year, *but **not** 20% of the original value!*

To find the value of the car after two years, subtract 20% of the value of the car after one year:

$$\$12,600 - (0.20)(\$12,600) = \$12,600 - \$2520$$

$$= \$10,080$$

To find the value of the car after three years, subtract 20% of the value of the car after two years:

$$\$10,080 - (0.20)(\$10,080) = \$10,080 - \$2016$$

$$= \$8064$$

$8064 is less than half of $18,000, so the car's value will decrease to less than half its value during the third year after its purchase.

c. Suppose the value of another car, which also costs $18,000, decreases at the rate of 25% each year. Which car would have the greater value 3 years after it was purchased? Explain or show how you found your answer.

To find the value of the second car after three years, use the method described in part b. (values are rounded to the nearest dollar):

the value of the car after one year $= \$18,000 - (0.25)(\$18,000) = \$18,000 - \$4500$

$$= \$13,500$$

the value of the car after two years $= \$13,500 - (0.25)(\$13,500) = \$13,500 - \$3375$

$$= \$10,125$$

the value of the car after three years $= \$10,125 - (0.25)(\$10,125) = \$10,125 - \$2531$

$$= \$7594$$

After three years, car 1 is worth more with a value of $8064.

Practice Problems

Name _____

Find the percent of change:

1) original number: 60
   new number: 45

2) original number: 120
   new number: 108

3) original number: 18
   new number: 24

4) original number: 10
   new number: 25

5) original number: 28
   new number: 14

6) original number: 30
   new number: 36

Find the original number:

7) new number: 24
   percent of change: 50%

8) new number: 85
   percent of change: −15%

9) new number: 160
   percent of change: 60%

10) new number: 76
    percent of change: −5%

11) new number: 228
    percent of change: 14%

12) new number: 42
    percent of change: −30%

Find the new number:

13) original number: 120   14) original number: 20   15) original number: 70
    percent of change: 60%   percent of change: 25%   percent of change: −30%

16) original number: 40   17) original number: 240   18) original number: 72
    percent of change: 10%   percent of change: −1%   percent of change: 25%

Convert each sentence to an equation and then solve:

19) What is 30% more than 60?

20) What number decreased by 35% is 65?

21) 24 is 20% more than what number?

22) An item sells for $88 after a 20% discount. Find the original price.

23) If the state sales tax is 5%, how much will you pay for an item marked $9.99?

Lesson 9: Factoring

Factors

A **factor** is an integer that divides evenly into another number.

example 1

The number 18 has a total of 6 factors: 1, 2, 3, 6, 9, and 18. What is the total number of factors that the number 130 has?

A. 4

B. 6

C. 8

D. 13

One important thing to understand about **factors** is that they come in *pairs*. The factors for 18 are given above as a list, but come from: 1 × 18, 2 × 9, and 3 × 6. Multiplying each pair of factors together equals 18.

To find all the factors of 130, you need to find each pair of numbers that can be multiplied together to get a product of 130. One way to do this is to use the opposite of multiplication, which is division. Start dividing 130 by 2, 3, 4, etc. to see if they divide evenly into 130. Your first factors will always be 1 and the number itself (1 × 130 = 130).

$130 \div 2 = 65$ So, $2 \times 65 = 130$ ⇒   2 and 65 are factors of 130

$130 \div 3 = 43.333...$ 3 does not divide evenly into 130, so it's not a factor.

$130 \div 4 = 32.5$ 4 also does not divide evenly into 130; 4 isn't a factor.

$130 \div 5 = 26$ So, $5 \times 26 = 130$ ⇒   5 and 26 are factors of 130

6, 7, 8, and 9 all do *not* divide evenly into 130; they are not factors.

$130 \div 10 = 13$ So, $10 \times 13 = 130$ ⇒   10 and 13 are factors of 130

11 and 12 do not divide evenly into 130.

We can stop checking numbers at this point because the next number to test is 13, but we've already found 13 as a factor. Notice how each quotient found got smaller and smaller (65, then 43.3, then 32.5, etc.) while you tested larger and larger integers (2, then 3, then 4, etc.). As soon as a quotient is found to be smaller than the number you are testing, you can stop searching for factors, because at this point you'll start repeating factors. (Such 10 and 13, and then 13 and 10).

The factors that were found were: 1 × 130, 2 × 65, 5 × 26, and 10 × 13. Or you can list them out as 1, 2, 5, 10, 13, 26, 65, and 130.

The answer is C.

Prime Numbers

A **prime number** is a whole number that has only one pair of factors: the number 1 and itself. This means a prime number cannot be divided evenly by any other whole numbers. (A **whole number** is a nonnegative integer: 0, 1, 2, 3, etc.)

The table below shows the first 100 prime numbers:

2	3	5	7	11	13	17	19	23	29
31	37	41	43	47	53	59	61	67	71
73	79	83	89	97	101	103	107	109	113
127	131	137	139	149	151	157	163	167	173
179	181	191	193	197	199	211	223	227	229
233	239	241	251	257	263	269	271	277	281
283	293	307	311	313	317	331	337	347	349
353	359	367	373	379	383	389	397	401	409
419	421	431	433	439	443	449	457	461	463
467	479	487	491	499	503	509	521	523	541

Composite Numbers

Except for 0 and 1, a whole number is either prime or composite. A **composite number** is any whole number, greater than one, that is not prime. This means a composite number has more than two factors, or in other words, it can be evenly divided by another whole number *besides* 1 and itself.

All even numbers, except for 2, are composite numbers. (This also means that 2 is the only even prime number, as shown in the table above.)

Greatest Common Factors

Two or more numbers can have the same factors, called common factors. For example, the factors of 20 are 1, 2, 4, 5, 10, and 20 and the factors of 30 are 1, 2, 3, 5, 6, 10, 15, and 30. The common factors of 20 and 30 are 1, 2, 5, and 10. The largest of the common factors is called the greatest common factor, which is 10 in this example.

To find the greatest common factor among two or more numbers, write out the **prime factorization** of each number using the following procedure:

Divide each number by the smallest prime number by which it can be divided. Start with 2 if the number is even, or 3 if the number is odd, and then go up one prime number at a time (5, 7, 11, etc.) until you have found all of the prime numbers that can be multiplied together to create the number you started with.

example 2

Write out the prime factorization of 140 and 72.

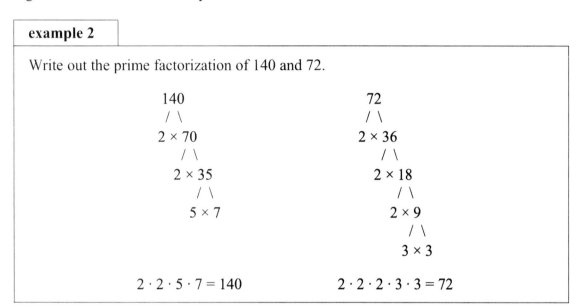

Once the prime factorization is written out, any common factors can be easily seen:

$$2 \cdot 2 \cdot 5 \cdot 7 \qquad\qquad 2 \cdot 2 \cdot 2 \cdot 3 \cdot 3$$

For the numbers 140 and 72, they each have prime factors of 2 and another 2, so multiply the common prime factors together, and a greatest common factor of 4 is found.

example 3

What is the prime factorization of 300?

A. $3 \cdot 10^2$
B. $2 \cdot 5^2 \cdot 6$
C. $2^2 \cdot 3 \cdot 25$
D. $2^2 \cdot 3 \cdot 5^2$

The prime factorization of 300 can be written out as:

$$300$$
$$/ \ \backslash$$
$$2 \times 150$$
$$/ \ \backslash$$
$$2 \times 75$$
$$/ \ \backslash$$
$$3 \times 25$$
$$/ \ \backslash$$
$$5 \times 5$$

$$2 \cdot 2 \cdot 3 \cdot 5 \cdot 5 = 300$$

or as: $2^2 \cdot 3 \cdot 5^5 = 300$

The answer is D.

Practice Problems

Name _____

Write out the prime factorization of each number.

1) 30

2) 108

3) 175

4) 210

5) 416

6) 630

7) 875

8) 2310

Find the greatest common factor of the numbers given.

9) 20, 50

10) 16, 56

11) 36, 60

12) 54, 81

13) 128, 256

14) 136, 168

15) 28, 32, 36

16) 45, 63, 81

Lesson 10: Ratios and Proportions

A **ratio** is a comparison of one number to another.

---

example 1

In a box of donuts, if there are 8 plain donuts and 4 jelly donuts, the ratio of plain donuts to jelly donuts can be written in the following ways:

$$8 \text{ to } 4 \qquad 8:4 \qquad \frac{8}{4}$$

Ratios can be reduced the same way that fractions can. The above ratios can be rewritten as:

$$2 \text{ to } 1 \qquad 2:1 \qquad \frac{2}{1}$$

Note: Because a ratio can be first presented in reduced form, it doesn't necessarily represent the actual number of items.

(When a ratio compares two quantities with different units of measure, that ratio is called a *rate*, and it is usually reduced to have a denominator of 1.)

A ratio is most commonly expressed as a fraction. This allows one ratio to be set equal to another ratio in what is called a **proportion**.

---

example 2

$$\frac{3}{6} = \frac{1}{2}$$

The ratio of 3 to 6 is equivalent (or *proportional*) to the ratio of 1 to 2.

To check if two ratios are proportional, cross-multiply. If the cross products are equal, then the ratios are proportional.

---

example 3

Are the given ratios proportional?

$$\frac{4}{7} \overset{?}{=} \frac{8}{15}$$

$$60 \overset{\nwarrow}{\underset{\nearrow}{\times}} 56$$
$$\frac{4}{7} \qquad \frac{8}{15}$$

$$60 \neq 56$$

So, $\dfrac{4}{7} \neq \dfrac{8}{15}$

---

example 4

Are the given ratios proportional?

$$\dfrac{3}{11} \overset{?}{=} \dfrac{9}{33}$$

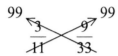

$$99 = 99$$

So, $\dfrac{3}{11} = \dfrac{9}{33}$

If an unknown variable is involved in one of two ratios set proportional to each other, the unknown can be found by following the same method of cross-multiplying.

example 5

Solve for *x*:

$$\dfrac{12}{9} = \dfrac{4}{x}$$

$$12x = 36$$

$$\dfrac{12x}{12} = \dfrac{36}{12}$$

$$x = 3$$

Cross-multiplying follows the same equality properties for solving equations and can sometimes save a step or two. The same problem from **example 5** is shown below, but without cross-multiplying:

example 6

Solve for x:

$$\frac{12}{9} = \frac{4}{x}$$

$$x \cdot \left(\frac{12}{9}\right) = \left(\frac{4}{x}\right) \cdot x$$

$$\frac{12x}{9} = 4$$

$$9 \cdot \left(\frac{12x}{9}\right) = 4 \cdot 9$$

$$12x = 36$$

$$\frac{12x}{12} = \frac{36}{12}$$

$$x = 3$$

Word problems can be solved using proportions as well.

example 7

In a recent year, there were 1682 students enrolled in Amherst College in Amherst, Massachusetts. The ratio of the number of students to the number of faculty members was approximately 15:2.

Based on this ratio, which of the following is closest to the number of faculty members that year?

A. 310
B. 220
C. 120
D. 110

A proportion can be set up where the ratio of 15 to 2 is approximately equal to the ratio of 1682 to x, which is the unknown number of faculty members:

$$\frac{15}{2} \approx \frac{1682}{x}$$

cross-multiply to get:

$$15x \approx 3364$$

$$\frac{15x}{15} \approx \frac{3364}{15}$$

$$x \approx 224$$

The answer is B.

---

example 8

If rope costs $2 per yard, how much would 75 feet of rope cost?

In this problem, first notice the change in units from yards to feet in the wording of the problem. The ratios in the proportion to be used must have the same unit of measurement.

$$\frac{\$2}{1 \text{ yard}} = \frac{\$x}{75 \text{ feet}}$$

$$\frac{\$2}{3 \text{ feet}} = \frac{\$x}{75 \text{ feet}}$$

Now that the unit measurements match, drop the units and cross multiply:

$$\frac{2}{3} = \frac{x}{75}$$

$$150 = 3x$$

$$\frac{150}{3} = \frac{3x}{3}$$

$$x = 50$$

75 feet of rope would cost $50.

Practice Problems

Name _____

Solve each proportion.

1) $\dfrac{x}{2} = \dfrac{3}{6}$

2) $\dfrac{-3}{4} = \dfrac{a}{12}$

3) $\dfrac{3}{7} = \dfrac{-18}{p}$

4) $\dfrac{d}{4} = \dfrac{9}{6}$

5) $\dfrac{5}{3} = \dfrac{b}{2}$

6) $\dfrac{7}{-2} = \dfrac{6}{n}$

7) $\dfrac{x+2}{12} = \dfrac{3}{4}$

8) $\dfrac{9}{5} = \dfrac{c-4}{5}$

9) $\dfrac{2-h}{6} = \dfrac{4}{3}$

10) $\dfrac{5}{m+1} = \dfrac{10}{4}$

Use a proportion to solve each word problem.

11) The ratio of boys to girls in a school is 3:2. If there are 600 male students, how many females students go to that school?

12) On average, a bag of 20 pieces of candy has 5 cherry-flavored pieces. How many bags of candy would be needed in order to have 25 cherry-flavored candies.

13) Steven mows lawns during the summer and makes $20 per lawn. How many lawns does he need to mow to earn $300?

14) A mailman can deliver mail to 3 houses every 5 minutes. How many houses will receive mail after 2 hours?

15) A fruit basket company puts 3 oranges, 4 bananas, and 5 apples in each basket. If 30 fruit baskets were ordered today, how many pieces of fruit would be needed?

16) A small forest contains only maple trees or pine trees. For every 3 maple trees, there are 8 pines trees. If there are 1100 trees in the forest, how many of them are maple trees?

Lesson 11: Direct and Inverse Variation

<u>Direct Variation</u>

Direct variation refers to a pair of variables related by what is called a **constant of variation**:

$$y = k \cdot x$$

x and y are the two variables
k is the constant of variation

---

example 1

If you work at a job that pays $8 per hour, the relationship between the number of hours worked (x) and the amount of income earned (y) can be expressed as:

$$y = 8x$$

where 8 is the constant of variation.

Direct variation problems often involve proportions.

---

example 2

If you earn $32 for working 4 hours, how much will you earn for working 12 hours?

The ratio of $32 to 4 gives a constant of variation (or *rate*) of $8 per hour. However, a proportion can be set up instead that doesn't make use of this constant of variation:

$$\frac{\$32}{4 \text{ hours}} = \frac{\$x}{12 \text{ hours}}$$

Now solve this proportion for x.

$$(12)\left(\frac{32}{4}\right) = \left(\frac{x}{12}\right)(12)$$

$$x = 96$$

You will earn $96 for working 12 hours if you would have made $32 for working 4 hours.

This direct variation proportion can be generically written as:

$$\frac{x_1}{y_1} = \frac{x_2}{y_2}$$ **formula for direct variation**

example 3

Assume x varies directly as y. If $x = 5$ when $y = 12$, find y when $x = 15$.

In this problem, $x_1 = 5$, $y_1 = 12$, $x_2 = 15$, and y_2 is the unknown. Plug these into the formula for direct variation:

$$\frac{x_1}{y_1} = \frac{x_2}{y_2}$$

$$\frac{5}{12} = \frac{15}{y_2}$$

Now solve for y by cross multiplying (and the subscript can be dropped at this point).

$$5y = 180$$

$$\frac{5y}{5} = \frac{180}{5}$$

$$y = 36$$

If $x = 5$ when $y = 12$, then $y = 36$ when $x = 15$.

Inverse Variation

Inverse variation refers to two variables that are inversely proportional to each other, meaning that as one variable increases in value, the other variable must decrease proportionally.

example 4

If you have a 120-mile trip to make, the faster you drive the less time it would take to complete the trip.

A table of rates and times can be made as follows:

rate (MPH)	20	30	40	60	80
time (hours)	6	4	3	2	1.5

The one value that remains constant is the distance, which can be computed by multiplying together the two variables in question (rate ? time).

For inverse variation, this is expressed as:

$$x \cdot y = k$$

x and y are the two variables
k is the constant of variation

In the previous example, you can see how one variable (rate) multiplied by the other variable (time) always results in a constant of variation (distance) of 120 miles.

Just as with direct variation, another formula can be used that does not use the constant of variation:

$$x_1 y_1 = x_2 y_2$$ **formula for inverse variation**

---

example 5

Assume x varies inversely as y. If $x = 4$ when $y = 6$, find y when $x = 12$.

In this problem, $x_1 = 4$, $y_1 = 6$, $x_2 = 12$, and y_2 is the unknown. Plug these into the above formula:

$$x_1 y_1 = x_2 y_2$$

$$(4)(6) = (12)y_2$$

Now solve for y (and drop the subscript).

$$24 = 12y$$

$$\frac{24}{12} = \frac{12y}{12}$$

$$y = 2$$

If $x = 4$ when $y = 6$, then $y = 2$ when $x = 12$.

THIS PAGE INTENTIONALLY BLANK

Practice Problems

Name _____

Solve each problem assuming that y varies directly as x.

1) If $y = 5$ when $x = 2$, what is y when
    $x = 10$?

2) If $y = 3$ when $x = -3$, what is x when
    $y = 6$?

3) If $y = 6$ when $x = 1$, what is y when
    $x = -2$?

4) If $y = -7$ when $x = 4$, what is x when
    $y = 21$?

5) If $y = 18$ when $x = 24$, what is y when
    $x = -36$?

6) If $y = 6$ when $x = -10$, what is x when
    $y = -3$?

7) If $y = -14$ when $x = 9$, what is y when
    $x = 6$?

8) If $y = 8$ when $x = 24$, what is x when
    $y = 3$?

Solve each problem assuming that y varies inversely as x.

9) If $y = 6$ when $x = 4$, what is y when
    $x = 3$?

10) If $y = -8$ when $x = -5$, what is x when
    $y = 10$?

11) If $y = 20$ when $x = 3$, what is y when
    $x = -15$?

12) If $y = -2$ when $x = 5$, what is x when
    $y = 1$?

13) If $y = 3$ when $x = 2$, what is y when
    $x = -3$?

14) If $y = 2$ when $x = -8$, what is x when
    $y = -4$?

15) If $y = 5$ when $x = 4$, what is y when
    $x = -6$?

16) If $y = 6$ when $x = 9$, what is x when
    $y = 2$?

Direct variation word problems.

17) If 8 gallons of gasoline cost $24.00, how much will 14 gallons cost?

18) A delivery service can deliver 5 packages every 2 hours. At that rate, how many packages can be delivered in 8 hours?

19) In the first 3 months of the year, the Jones family spent $2300 on groceries. What amount can they expect to spend on groceries for the rest of the year?

20) At a local carnival, Bill spent $3.00 to go on 4 rides. He wants to go on 14 rides before going home. How much will he end up spending at the carnival?

Inverse variation word problems.

21) Mr. Jones has to go on a business trip. If he can average 60 miles per hour while driving there, it will take him 4 hours. How long will it take if he averages 40 miles per hour during his drive?

22) Marci weighs 120 pounds and is sitting 8 feet from the center (fulcrum) of a see-saw. If Jenn weighs 96 pounds, how many feet from the center does she need to sit in order to balance the see-saw?

23) If 3 workers can get a job done in 8 hours, how long would it take 4 workers to complete the same job?

24) At 6 gallons per minute, a hose can fill a water tank in 20 minutes. How long would it take to fill the same tank at 8 gallons per minute?

National Council of Teachers of Mathematics
Standards for School Mathematics

Algebra Standard for Grades 6–8

Expectations

Instructional programs from prekindergarten through grade 12 should enable all students to—	In grades 6–8 all students should—	The following lessons correspond to each expectation—
Understand patterns, relations, and functions	• represent, analyze, and generalize a variety of patterns with tables, graphs, words, and, when possible, symbolic rules; • relate and compare different forms of representation for a relationship; • identify functions as linear or nonlinear and contrast their properties from tables, graphs, or equations.	• Lessons 12, 13, and 14 • Lesson 15 • Lessons 15 and 28
Represent and analyze mathematical situations and structures using algebraic symbols	• develop an initial conceptual understanding of different uses of variables; • explore relationships between symbolic expressions and graphs of lines, paying particular attention to the meaning of intercept and slope; • use symbolic algebra to represent situations and to solve problems, especially those that involve linear relationships; • recognize and generate equivalent forms for simple algebraic expressions and solve linear equations.	• Lesson 30 • Lessons 15, 16, 17, and 18 • Lesson 30 • Lessons 20, 21, and 30
Use mathematical models to represent and understand quantitative relationships	• model and solve contextualized problems using various representations, such as graphs, tables, and equations.	• Lessons 18 and 30
Analyze change in various contexts	• use graphs to analyze the nature of changes in quantities in linear relationships.	• Lesson 18

Lesson 12: Arithmetic and Geometric Sequences

A **sequence** is an ordered set of numbers that can be either finite or infinite.

A **finite sequence** has a finite number of terms, while an **infinite sequence** goes on forever, or in other words, it has an infinite number of terms.

Each **term** of a sequence is determined by a **function** or **formula** (also called a **recursion formula**).

Arithmetic Sequences

example 1
The sequence 1, 4, 7, 10, 13, … is created by taking the first term and adding three over and over.

The notation used for sequences is different than anything you may have seen before.

a is typically used to represent each term, with subscripts to identify *which* term of the sequence represents. So for 1, 4, 7, 10, 13, …

$a_1 = 1$, $a_2 = 4$, $a_3 = 7$, $a_4 = 10$, **and so on.**

a_n is used to represent the nth term of a sequence.

In the example above,

$$a_n = 1 + (n-1) \cdot 3$$

1 is the first term of the sequence

3 is the difference between each term

Any sequence that is defined by *adding* or *subtracting* the same number over and over is called an **arithmetic sequence**. The following recursion formula can be used for any arithmetic sequence:

$$a_n = a_1 + (n-1) \cdot d$$

where a_1 is the first term of the sequence and d is the difference (called the **common difference**) between each term.

example 2

What is the 9th term of the following arithmetic sequence:

32, 29, 26, …

$a_1 = 32$

$d = -3$

a_9 is the 9th term, so $n = 9$

Using the formula for an arithmetic sequence:

$a_n = a_1 + (n - 1) \cdot d$

$a_9 = 32 + (9 - 1)(-3)$

$a_9 = 32 + (8)(-3)$

$a_9 = 32 + (-24)$

$a_9 = 8$

The 9th term is 8.

If you can't remember this formula, you can always write out all of the terms by adding d over and over until you get to the nth term.

$$32, \quad 29, \quad 26, \quad 23, \quad 20, \quad 17, \quad 14, \quad 11, \quad 8$$
$$\uparrow \quad \uparrow \quad \uparrow \quad \uparrow \quad \uparrow \quad \uparrow \quad \uparrow \quad \uparrow \quad \uparrow$$
$$a_1 \quad a_2 \quad a_3 \quad a_4 \quad a_5 \quad a_6 \quad a_7 \quad a_8 \quad a_9$$

Geometric Sequences

When a sequence is created by starting with some first term and then *multiplying* or *dividing* by the same number over and over again, it is called a **geometric sequence**.

The recursion formula for a geometric sequence is:

$$a_n = a_1 r^{n-1}$$

where a_1 is the first term of the sequence, but instead of a constant difference d between terms, a **common ratio** r is used. r is the number that the first term is multiplied by over and over to create the sequence. (r is also the ratio of any term of the sequence to any previous term of the sequence: $\dfrac{a_2}{a_1} = r$, $\dfrac{a_5}{a_4} = r$, etc.)

---

example 3

What is the 10th term of the following geometric sequence:

  2, 4, 8, 16, 32, …

In this sequence, the first term is multiplied by 2, and then the resulting term is multiplied by 2 to create the next term, and so on, so r must be 2. (r can also be found by finding the ratio of any two consecutive terms of the sequence: $\frac{4}{2} = 2$, $\frac{8}{4} = 2$, $\frac{16}{8} = 2$, etc.)

$$a_1 = 2$$

$$r = 2$$

$$n = 10$$

Using the formula for a geometric sequence:

$$a_n = a_1 r^{n-1}$$

$$a_{10} = (2)(2)^{10-1}$$

$$a_{10} = (2)(2)^9$$

$$a_{10} = (2)(512)$$

$$a_{10} = 1024$$

The 10th term of this sequence is 1024.

---

example 4

The first number in a pattern is 50. To go from one number in the pattern to the next number, the rule is to **divide by 5**. What is the fourth number in the pattern?

A. $\frac{1}{5}$ C. $\frac{3}{2}$

B. $\frac{2}{5}$ D. $\frac{5}{2}$

One way to find the answer is to divide 50 by 5 to find the second number in the pattern, and then divide *that* number by 5 to find the third number, and so on:

1st term: 50

2nd term: $50 \div 5 = 10$

3rd term: $10 \div 5 = 2$

4th term: $2 \div 5 = \dfrac{2}{5}$

The answer is B.

Another method for finding the answer is to use a recursion formula. Because each term is being *divided* over and over, this is a *geometric* sequence. The ratio from one term to the next is $\dfrac{1}{5}$ (which is the same as dividing by 5).

Using the formula for a geometric sequence:

$$a_n = a_1 r^{n-1}$$

$$a_4 = (50)\left(\frac{1}{5}\right)^{4-1}$$

$$a_4 = (50)\left(\frac{1}{5}\right)^{3}$$

$$a_4 = (50)\left(\frac{1}{5^3}\right)$$

$$a_4 = \frac{50}{125} = \frac{25 \times 2}{25 \times 5} = \frac{2}{5}$$

The 4th term of this sequence is $\dfrac{2}{5}$.

Once again we see the answer is B.

Practice Problems

Name _____

Determine the common difference or common ratio to solve.

1) Find the next two terms of the sequence:

6, 12, 18, 24, _____, _____, …

2) Find the next three terms of the sequence:

2, 4, 8, 16, 32, _____, _____, _____, …

3) Find the next four terms of the sequence:

35, 30, 25, 20, _____, _____, _____, _____, …

4) Find the next two terms of the sequence:

4000, −2000, 1000, −500, _____, _____, …

Use a recursion formula to solve.

5) Find the 12th term of the sequence:

8, 14, 20, 26, …

6) Find the 7th term of the sequence:

1, 4, 16, 64, …

7) Find the 30th term of the sequence:

−9, −5, −1, 3, …

8) Find the 14th term of the sequence:

1, 2, 4, 8, …

Use the given arithmetic or geometric recursion formula to solve.

9) Find the 6th term of: 10) Find the 12th term of:

$$a_n = 3 + (n - 1) \cdot 6$$ $$a_n = 3(2)^{n-1}$$

11) Find the 8th term of: 12) Find the 11th term of:

$$a_n = 54 + (n - 1) \cdot (-3)$$ $$a_n = 768\left(\frac{1}{2}\right)^{n-1}$$

13) Find the 14th term of: 14) Find the 6th term of:

$$a_n = -4 + (n - 1) \cdot (2)$$ $$a_n = 50\left(\frac{1}{5}\right)^{n-1}$$

Find a recursion formula for the given arithmetic or geometric sequence.

15) 8, 14, 20, 26, 32, …   16) 4, 12, 36, 108, …

17) 85, 80, 75, 70, 65, …   18) 1944, 648, 216, …

Lesson 13: Fibonacci and Other Sequences

The Fibonacci Sequence

Not all sequences are arithmetic or geometric. Another type of sequence is named after the mathematician that discovered it:

$$0, 1, 1, 2, 3, 5, 8, 13, 21, 34, 55, \ldots$$

This is called the **Fibonacci sequence**. Each term (except the first two terms) is the sum of the previous two terms.

There are an infinite number of other sequences that can be created besides arithmetic and geometric sequences, and the Fibonacci sequence is just one example.

Ask yourself: Could you have figured out the pattern followed in the Fibonacci sequence if it wasn't already explained?

It is easy to make other Fibonacci-like sequences by starting with any two numbers, adding them together to create the next term, and then continuing in the same pattern to create additional terms, such as:

$$3, 4, 7, 11, 18, 29, 47, \ldots$$

Other Sequences

example 1

Find the next three terms by using the pattern found in the following sequence:

$$10, 9, 12, 11, 14, 13, 16, 15, 18, \ldots$$

This sequence is created by subtracting one, then adding three, over and over.

So if 1 is subtracted from 18, the next term is 17, and if 3 is then added to that, the following term is 20. Subtract 1 again, and the next term is 19:

$$10, 9, 12, 11, 14, 13, 16, 15, 18, \mathbf{17, 20, 19}, \ldots$$

---

example 2

What number comes next in this sequence?

$$5, 6, 9, 14, 21, \underline{\hspace{1cm}}$$

A. 26
B. 27
C. 30
D. 32

In this sequence, the terms increase according to the following pattern:

$$5, \overset{+1}{\nearrow\searrow} 6, \overset{+3}{\nearrow\searrow} 9, \overset{+5}{\nearrow\searrow} 14, \overset{+7}{\nearrow\searrow} 21, \ldots$$

Each term is increased according to a sequence of odd numbers (1, 3, 5, 7, …). So the next term in this sequence will come from adding 9 to the fifth term:

$$5, \overset{+1}{\nearrow\searrow} 6, \overset{+3}{\nearrow\searrow} 9, \overset{+5}{\nearrow\searrow} 14, \overset{+7}{\nearrow\searrow} 21, \overset{+9}{\nearrow\searrow} 30, \ldots$$

The answer is C.

---

example 3

The Pizza Palace's price list for plain pizza is shown below.

DIAMETER	COST
10"	$5.00
12"	$7.20
14"	$9.80
16"	$12.80

Based on this information, what would a 20" pizza likely cost?

A. $10.00
B. $14.40
C. $14.80
D. $20.00

In this problem, notice the pattern in how the cost of each larger pizza increases:

$5.00
+ $2.20   to $7.20
+ $2.60   to $9.80
+ $3.00   to $12.80

Each price increase *increases* by 40 cents.

So for the price of an 18" pizza, add $3.40 to get $16.20.

To get the price of a 20" pizza, add $3.80 to get a price of $20.00.

The answer is D.

Determining a Pattern Rule

example 4

n	1	2	3	4	5	6
a_n	0	3	8	15	24	35

If the pattern in the table continues, which of the following expressions represents a_n?

A. $2n - 1$
B. $(n - 1)^2$
C. $3(n - 1)$
D. $n^2 - 1$

The best approach to this problem is to just try each answer and see which one works. This is one of the advantages to a multiple choice question – that four possible answers are already given for you to check.

A) $a_n = 2n - 1$

        if $n = 1$, $a_n = 2(1) - 1 = 2 - 1 = 1$

        but a_n should equal 0 when $n = 1$

B) $a_n = (n - 1)^2$

        if $n = 1$, $a_n = (1 - 1)^2 = (0)^2 = 0$

        if $n = 2$, $a_n = (2 - 1)^2 = (1)^2 = 1$

        but a_n should equal 3 when $n = 2$

C) $a_n = 3(n - 1)$

        if $n = 1$, $a_n = 3(1 - 1) = 3(0) = 0$

        if $n = 2$, $a_n = 3(2 - 1) = 3(1) = 3$

        if $n = 3$, $a_n = 3(3 - 1) = 3(2) = 6$

        but a_n should equal 8 when $n = 3$

To this point, none of the answers have worked, and notice that sometimes *several* values had to be checked before a possible answer turned out to be wrong.

D) $a_n = n^2 - 1$

$$\text{if } n = 1, a_n = 1^2 - 1 = 1 - 1 = 0$$

$$\text{if } n = 2, a_n = 2^2 - 1 = 4 - 1 = 3$$

$$\text{if } n = 3, a_n = 3^2 - 1 = 9 - 1 = 8$$

$$\text{if } n = 4, a_n = 4^2 - 1 = 16 - 1 = 15$$

$$\text{if } n = 5, a_n = 5^2 - 1 = 25 - 1 = 24$$

$$\text{if } n = 6, a_n = 6^2 - 1 = 36 - 1 = 35$$

The answer is D.

---

example 5

Which equation states a rule for the pattern shown in the table below?

Input (x)	1	2	3	4
Output (y)	1	5	11	19

A. $y = x^2 - x + 1$
B. $y = x^2 + x - 1$
C. $y = x^2 + 3$
D. $y = x^2 + 1$

A sequence is sometimes determined from a rule that can be expressed as an equation. If y is each term in the sequence (a_n), and x is the position of each term in the sequence (n), an equation can be found to represent the pattern rule for this sequence.

Luckily, we don't have to find this equation on our own, as four possible answers have been provided. All you need to do is check each one and see which one works. Try each value of x and see if it matches the corresponding value for y in the table:

Answer A:

$$\text{when } x = 1, y = (1)^2 - (1) + 1 = 1 - 1 + 1 = 1$$
$$\text{when } x = 2, y = (2)^2 - (2) + 1 = 4 - 2 + 1 = 3$$
y should equal 5 when $x = 2$, so the answer is not A.

Answer B:

$$\text{when } x = 1, y = (1)^2 + (1) - 1 = 1 + 1 - 1 = 1$$
$$\text{when } x = 2, y = (2)^2 + (2) - 1 = 4 + 2 - 1 = 5$$
$$\text{when } x = 3, y = (3)^2 + (3) - 1 = 9 + 3 - 1 = 11$$
$$\text{when } x = 4, y = (4)^2 + (4) - 1 = 16 + 4 - 1 = 19$$

All of these x- and y-values match the table given in the problem, so the answer is B.

You should continue and check Answers C and D to make sure those answers do not work just in case you made a mistake finding that the answer is B.

Answer C:

$$\text{when } x = 1, y = (1)^2 + 3 = 1 + 3 = 4$$
y should equal 1 when $x = 1$, so the answer is not C.

Answer D:

$$\text{when } x = 1, y = (1)^2 + 1 = 1 + 1 = 2$$
y should equal 1 when $x = 1$, so the answer is not D.

---

example 6

Write a rule that could be used to show the relationship between x and y in the table below.

x	y
−4	16
−1	1
0	0
3	9
7	49

This problem is different from the previous example in two ways: 1) We don't have four possible answers from which to choose, and 2) the x-values don't correspond to the position of each term in the sequence (such as $x = 1$, $x = 2$, etc.). However, we still need to find some kind of relationship between the x-values and y-values that can be expressed as an equation.

The thing to notice in this problem is that for each value of x, y equals that value squared. [$16 = (-4)^2$, $1 = (-1)^2$, $0 = (0)^2$, $9 = (3)^2$, and $49 = (7)^2$]

This can be expressed as the following equation:

$$y = x^2$$

THIS PAGE INTENTIONALLY BLANK

Practice Problems

Name _____

Find the next two terms of each sequence.

1) 0, 1, 1, 2, 3, 5, 8, 13, 21, 34, 55, _____, _____, …

2) 2, 2, 4, 6, 10, 16, 26, 42, _____, _____, …

3) 3, 4, 7, 11, 18, 29, 47, 76, _____, _____, …

4) 3, 4, 6, 9, 13, 18, 24, 31, _____, _____, …

5) 30, 29, 27, 24, 20, 15, _____, _____, …

6) 5, 9, 8, 12, 11, 15, 14, 18, 17, _____, _____, …

7) 4, 7, 9, 12, 14, 17, 19, _____, _____, …

8) 1, 2, 6, 15, 31, 56, 92, _____, _____, …

9) 1, 3, 6, 11, 18, 29, 42, _____, _____, …

10) 4, 2, 3, 1, 4, 2, 7, 5, 12, _____, _____, …

11) Which equation states a rule for the pattern shown in the table?

x	1	2	3	4
y	2	4	8	14

A) $y = x^2 + 1$

B) $y = x^2 + x - 1$

C) $y = x^2 + 2$

D) $y = x^2 - x + 2$

12) Which equation states a rule for the pattern shown in the table?

x	0	1	2	3
y	3	4	7	12

A) $y = x^2 - x + 3$

B) $y = x + 3$

C) $y = x^2 + 3$

D) $y = x^2 + x + 2$

Lesson 14: Patterns — Numbers, Shapes, etc.

Patterns can be found in other things besides number sequences, such as a series of shapes like in **example 1** below.

A pattern can also be a list of numbers that is not a sequence:

$$2, 3, 7, 1, 4, 2, 3, 7, 1, 4, 2, 3, 7, 1, 4, \ldots$$

This list of numbers is not a sequence, it is a pattern of five numbers that repeats over and over. To determine some future term, use either of the two methods shown in the following two examples.

---

example 1

What would be the 25th shape in the pattern?

$$\square, O, \diamond, \square, O, \diamond, \ldots$$

In this pattern, a square, circle, and diamond repeat over and over. Since the pattern repeats after every third shape, we know that every term that is a multiple of three will be the third shape ($\diamond$). The closest multiple of three to 25 is 24.

Since the 24th term will be a diamond, the 25th term will be a square.

---

example 2

The first eight positions in a pattern are shown below.

$$N, S, E, W, N, S, E, W, \ldots$$

If this pattern continues, which letter would be found at the 103rd position?

A. E
B. S
C. N
D. W

Since the pattern repeats every 4th term, another approach to solve a problem like this is to divide 103 by 4 and see what the remainder is.

term 1	term 2	term 3	term 4
N	S	E	W

When 103 is divided by 4, you get a remainder of 3:

$$4\overline{)103} \quad 25R3$$

or $\quad \dfrac{103}{4} = 25\dfrac{3}{4}$

A remainder of 3 means the 103rd term is the same as the 3rd term, which is the letter E.

The answer is A.

Note: If there is no remainder when you divide, the answer is the same as the number you divided by. So with four terms, no remainder would mean the term you are looking for is the same as the 4th term, or the nth term if there are n terms in the pattern.

example 3

Each arrangement in the pattern below is made up of square tiles.

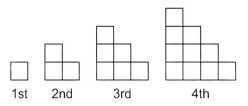

1st   2nd   3rd   4th

Which expression tells how many tiles are in the nth arrangement of this pattern?

A. $n(n + 1)$

B. $n(n - 1)$

C. $2n - 1$

D. $\dfrac{n}{2}(n + 1)$

The best way to find the correct answer here is to check each one. This is one of the benefits of a multiple choice question because the answer *must* be one of the four given.

You can see in the pattern of tiles that the first one has 1 tile, the second one has 3 tiles, the third one has 6 tiles, and the fourth one has 10 tiles. So, check each expression to see which one also produces the same pattern of numbers.

Investigating each answer:

Answer A: $n(n + 1)$

If $n = 1$, $n(n + 1) = 1(1 + 1) = 1(2) = 2$

But, there is 1 tile in the 1st arrangement, not 2, so A can't be the answer.

Answer B: $n(n - 1)$

If $n = 1$, $n(n - 1) = 1(1 - 1) = 1(0) = 0$

But, there is 1 tile in the 1st arrangement, not 0, so B can't be the answer.

Answer C: $2n - 1$

If $n = 1$, $2n - 1 = 2(1) - 1 = 2 - 1 = 1$
If $n = 2$, $2n - 1 = 2(2) - 1 = 4 - 1 = 3$
If $n = 3$, $2n - 1 = 2(3) - 1 = 6 - 1 = 5$

But, there are 6 tiles in the 3rd arrangement, not 5, so C can't be the answer.

Answer D: $\dfrac{n}{2}(n + 1)$

If $n = 1$, $\dfrac{n}{2}(n + 1) = \dfrac{1}{2}(1 + 1) = \dfrac{1}{2}(2) = 1$

If $n = 2$, $\dfrac{n}{2}(n + 1) = \dfrac{2}{2}(2 + 1) = \dfrac{2}{2}(3) = 3$

If $n = 3$, $\dfrac{n}{2}(n + 1) = \dfrac{3}{2}(3 + 1) = \dfrac{3}{2}(4) = 6$

If $n = 4$, $\dfrac{n}{2}(n + 1) = \dfrac{4}{2}(4 + 1) = \dfrac{4}{2}(5) = 10$

These all match the pattern of tiles.

The answer is D.

| example 4 |

The following shows the first five rows of Pascal's triangle.

Row 1:   1
Row 2:   1   1
Row 3:   1   2   1
Row 4:   1   3   3   1
Row 5:   1   4   6   4   1

Which of the following represents the **8th row**?

A. 1   7   21   35   35   21   7   1
B. 1   7   21   35   21   7   1
C. 1   5   5   10   10   5   5   1
D. 1   5   10   10   5   1

Pascal's triangle is a pattern formed by starting with a 1 in the first row, then two 1's in the second row underneath, with each 1 offset to the left or right of the 1 in the first row. Each successive row is made by starting with a 1, then each next term is found by adding the number above to the left and the number above to the right. When there are no more numbers above to the right, end that row with another 1.

Looking at the first five rows of Pascal's triangle given above, you should be able to see how Rows 3 through 5 are created from the numbers above them. Row 3 starts with a 1, and then the next term has a 1 above it to the left and a 1 above it to the right. Adding those together creates the next term in Row 3, which is a 2. For the next term, there is a 1 above it to the left, but no number above to the right, so the row ends with a 1. Row 4 starts with a 1, the next term is 3, which comes from adding the 1 above it to the left and the 2 that is above it to the right. And so on…

To answer this question, continue creating the triangle by creating Row 6, Row 7, and then Row 8:

Row 6:   1   5   10   10   5   1
Row 7:   1   6   15   20   15   6   1
Row 8: 1   7   21   35   35   21   7   1

The answer is A.

Practice Problems

Name _____

1) The first six rows of Pascal's triangle are shown below:

Row 1:   1
Row 2:   1   1
Row 3:   1   2   1
Row 4:   1   3   3   1
Row 5:   1   4   6   4   1
Row 6:   1   5   10   10   5   1

If the pattern continues, what is the sequence of numbers in the 9th row?

2) Given the following:

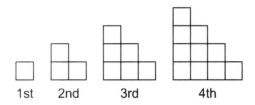

1st   2nd   3rd   4th

How many squares will be in the 12th arrangement of this pattern?

3) Given the following pattern:

If this pattern continues, in what direction will the arrow in the 18th position point (the first arrow points down, the second arrow points to the right, etc.)?

4) Given the following pattern:

   CIRCLE, SQUARE, TRIANGLE, RECTANGLE, CIRCLE, SQUARE, …

   If this pattern continues, what word will be in the 25th position?

5) Given the following pattern:

   If this pattern continues, what number (⓪,①, or ②) is in the 35th position?

Give the next two items for pattern shown.

6) x, $2x^2$, $3x^3$, $4x^4$, _____, _____, …

7) ○, ●, □, ■, ○, _____, _____, …

8) α, β, χ, δ, α, β, _____, _____, …

9) $a + 1$, $a + 4$, $a + 9$, $a + 16$, $a + 25$, _____, _____, …

10) Z1, Y2, X3, W4, _____, _____, …

Lesson 15: Linear Relationships

While a sequence formula or pattern rule can determine the relationship among the terms of a sequence or the terms of a pattern, a relationship can exist among **ordered pairs** (two numbers paired together) that can be plotted on a graph as a **line**. When such a **linear relationship** exists, the formula that determines this relationship is called a **linear equation**.

For an equation to be linear, all variables must be raised to the first power. There can't be any variables that are squared, cubed, etc., and there can't be any square roots either (or cube roots, etc). If you see any exponents or radicals ($\sqrt{}$), then the equation is *not* linear.

---

example 1

Which of the following equations does **not** represent a linear relationship?

A. $xy = 12$
B. $x + y = 12$
C. $y = 12x$
D. $x - y = 12$

A linear relationship between ordered pairs can be represented by a linear equation of the form: $y = m \cdot x + b$, where x and y are variables and m and b are rational numbers.

The answer to this question will be the one that *cannot* be rewritten in this form. In answer A, because x and y are multiplied together, x can only be moved to the other side of the equation by dividing both sides by x:

$$\frac{xy}{x} = \frac{12}{x}$$

$$y = \frac{12}{x}$$

This is definitely not of the form $y = m \cdot x + b$ because x is in a denominator.

The answer is A.

Determining the Rule for a Linear Relationship

example 2

x	1	2	3	4
y	4	7	10	13

What equation would represent the linear relationship between x and y given the values shown in the table above?

The first step is to determine the ratio of how y changes to how x changes. This ratio can be called m:

$$m = \frac{\text{change in } y}{\text{change in } x}$$

x: $1 \nearrow^{+1} 2 \nearrow^{+1} 3 \nearrow^{+1} 4$

y: $4 \searrow_{+3} 7 \searrow_{+3} 10 \searrow_{+3} 13$

The value of y changes by 3, while the value of x changes by 1.

So, $m = \dfrac{3}{1}$ or just 3.

To finish determining the equation that relates x and y together, use the equation:

$y = m \cdot x + b$ (form of a linear equation)

or in this case:

$y = 3 \cdot x + b$ since $m = 3$

To find c, substitute an ordered pair (a matching pair of x- and y-values) for x and y, and then solve for c:

$x = 1$ when $y = 4$:

$y = 3 \cdot x + b$

$4 = 3(1) + b$

$4 = 3 + b$
$-3 \quad -3$

$b = 1$

Substitute this value for b:

$$y = 3 \cdot x + b$$

$$y = 3 \cdot x + 1$$

$$y = 3x + 1$$

example 3

Write the rule for the table shown below.

Input (x)	Output (y)
3	5
6	11
2	3
8	15

In this example, the values for x and y increase, then decrease, and then increase again. This doesn't affect how the ratio of the change in y to the change in x is found:

Here, when the change in y is 6, the change in x is 3, when the change in y is −8, the change in x is −4, and when the change in y is 12, the change in x is 6:

$$m = \frac{\Delta y}{\Delta x} = \frac{6}{3} \text{ or } \frac{-8}{-4} \text{ or } \frac{12}{6}, \text{ all of which equal 2.}$$

(Note: Δ is the mathematical symbol for change, so Δy means "change in y".)

This value for m can now be used in the formula for any linear equation:

$$y = m \cdot x + b$$

or in this case:

$$y = 2 \cdot x + b$$

To find b, substitute an ordered pair (a matching pair of x- and y-values) for x and y, and then solve for b:

$x = 3$ when $y = 5$:

$y = 2 \cdot x + b$

$5 = 2(3) + b$

$5 = 6 + b$

$\underline{-6 \quad -6}$

$b = -1$

Substitute this value for b:

$y = 2 \cdot x + b$

$y = 2 \cdot x - 1$

$y = 2x - 1$

Practice Problems

Name _____

Label each equation as *linear* or *not linear*.

1) $x + y^2 = 3$

2) $y = x^3 - 5$

3) $2x - 3y = -5$

4) $y = 4x + 9$

5) $x^2 + y^2 = 9$

6) $x = 3$

7) $y = \sqrt{x - 3}$

8) $x + 4y = -7$

9) $y = 0$

10) $y = x^2 + 3$

11) $x^2 + 3x + 2 = 0$

12) $\dfrac{1}{2}x + \dfrac{3}{5}y = 4$

Write a linear equation for each set of ordered pairs.

13)

x	0	1	2	3	4	5
y	8	5	2	−1	−4	−7

14)

x	0	1	2	3	4	5
y	−2	1	4	7	10	13

15)

x	−2	−1	0	1	2	3
y	1	3	5	7	9	11

16)

x	−4	−2	0	2	4	6
y	4	3	2	1	0	−1

Lesson 16: Slope

Slope refers to a property of a line on a graph, specifically the ratio of the "rise" of the line to the "run" of the line. In other words, the slope of the line refers to the *steepness* of the line.

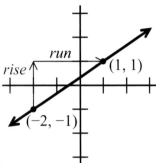

By comparing two points on the line, the "rise" (vertical change) and "run" (horizontal change) can be calculated.

"Rise" can be computed by subtracting the y-value of one point (y_1) from the y-value of a second point (y_2), and the "run" can be computed by subtracting the x-value of the first point (x_1) from the x-value of a second point (x_2).

$$\textbf{slope} = \frac{"rise"}{"run"} = \frac{y_2 - y_1}{x_2 - x_1}$$

If $x_2 = x_1$, you get a fraction with a zero in the denominator, *which is never allowed.* (You cannot divide by zero, so a zero is never allowed in the denominator.)

In this situation, the line has *no slope*, which also means it is a *vertical line.*

---

example 1

(2, 2)

(2, −1)

$$\text{slope } = \frac{y_2 - y_1}{x_2 - x_1} = \frac{2 - (-1)}{2 - 2}$$

$$= \frac{2 + 1}{0}$$

$$= \frac{3}{0} \Rightarrow \text{NO SLOPE}$$

If $y_2 = y_1$, the slope will have a value of *zero*, which means the line is *horizontal*.

example 2

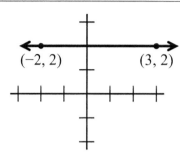

$$\text{slope} = \frac{y_2 - y_1}{x_2 - x_1} = \frac{2-2}{3-(-2)}$$

$$= \frac{0}{3+2}$$

$$= \frac{0}{5}$$

$$= 0$$

A line that ascends from left to right has a **positive slope**, while a line that descends from left to right has a **negative slope**.

example 3

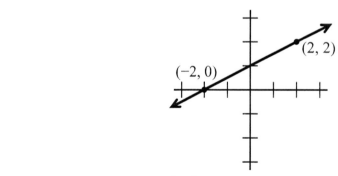

$$\text{slope} = \frac{y_2 - y_1}{x_2 - x_1} = \frac{2-0}{2-(-2)}$$

$$= \frac{2}{2+2}$$

$$= \frac{2}{4} \text{ or } \frac{1}{2} \Rightarrow \text{positive slope}$$

example 4

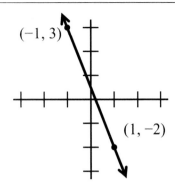

$$\text{slope} = \frac{y_2 - y_1}{x_2 - x_1} = \frac{-2 - 3}{1 - (-1)}$$

$$= \frac{-5}{1 + 1}$$

$$= \frac{-5}{2} \Rightarrow \text{negative slope}$$

example 5

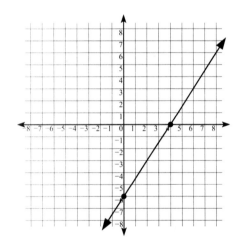

Which statement best describes the slope of the line graphed above?

A. The slope is –6.   C. The slope is $\frac{3}{2}$.

B. The slope is $-\frac{2}{3}$.   D. The slope is 4.

To calculate the slope of the line on the graph, first pick any two points on that line. The **x-intercept** (the point where the line crosses the x-axis) and the **y-intercept** (the point where the line crosses the y-axis) are indicated on the graph by the two points shown on the line. The x-intercept is point (4, 0) and the y-intercept is (0, −6).

Use these two points for the two x-values and two y-values needed to calculate slope:

$$\text{slope} = \frac{y_2 - y_1}{x_2 - x_1}$$

$$= \frac{-6 - 0}{0 - 4}$$

$$= \frac{-6}{-4}$$

The negative signs in the numerator and denominator cancel, and the fraction can be reduced:

$$\text{slope} = \frac{3}{2}$$

The answer is C.

---

example 6

Determine the value of c so that a line passes through points (2, 5), (−3, c), and has a slope of 1.

$$\text{slope} = \frac{y_2 - y_1}{x_2 - x_1}$$

$$1 = \frac{c - 5}{-3 - 2}$$

$$1 = \frac{c - 5}{-5}$$

$$(-5) \times 1 = \frac{c - 5}{-5} \times (-5)$$

$$-5 = c - 5$$
$$+5 \qquad +5$$

$$c = 0$$

Practice Problems

Name _____

Find the slope of the line that passes through the given pair of points.

1) (2, 2), (4, 10)   2) (2, 4), (5, 4)

3) (0, 2), (4, 5)   4) (3, 1), (3, 6)

5) (1, 1), (5, 9)   6) (5, 0), (8, 0)

7) (2, 3), (1, 5)   8) (−2, −3), (4, 2)

Find the value of c so that a line passing through the given points has the given slope.

9) $(2, c), (3, 5), m = 0$

10) $(3, c), (4, 9), m = 1$

11) $(7, 2), (c, 6), m = 4$

12) $(3, 3), (-4, c), m = 0$

13) $(c, 3), (2, -6)$, no slope

14) $(4, c), (8, 7), m = \dfrac{1}{2}$

Lesson 17: Equation of a Line

Starting with the equation for the slope of a line:

$$m = \frac{y_2 - y_1}{x_2 - x_1}$$

This can be rearranged to represent the equation of a line.

First drop the subscripts from the second point (x_2, y_2), then rearrange terms by multiplying each side by $(x - x_1)$:

$$m = \frac{y_2 - y_1}{x_2 - x_1} \Rightarrow m = \frac{y - y_1}{x - x_1}$$

$$(x - x_1) \cdot m = \frac{y - y_1}{x - x_1} \cdot (x - x_1)$$

$$(x - x_1) \cdot m = y - y_1$$

OR

$$y - y_1 = m(x - x_1)$$

This is called the **point-slope equation** of a line.

x_1 and y_1 are an ordered pair, which represents a point on a line, and m is the slope of that line.

example 1

Write the point-slope form of an equation of a line that passes through point (4, 3) and has a slope of $\frac{2}{3}$.

$$y - y_1 = m(x - x_1)$$

$$x_1 = 4$$

$$y_1 = 3$$

$$m = \frac{2}{3}$$

$$y - 3 = \frac{2}{3}(x - 4)$$

To write this answer in **standard form**, the equation must be rewritten in the form $Ax + By = C$, where A, B, and C are integers, A and B are not zero, and A is positive.

The previous equation written in point-slope form can be written in standard form as follows:

$$y - 3 = \frac{2}{3}(x - 4)$$

$$3 \cdot (y - 3) = \frac{2}{3}(x - 4) \cdot 3$$

$$3(y - 3) = 2(x - 4)$$

$$3y - 9 = 2x - 8$$
$$\underline{-2x \qquad -2x}$$

$$-2x + 3y - 9 = -8$$
$$\underline{\qquad +9 \quad +9}$$

$$-2x + 3y = 1$$

$$-1 \cdot (-2x + 3y) = 1 \cdot (-1)$$

$$2x - 3y = -1$$

Starting from the point-slope equation, another common type of equation of a line can be derived:

> If a line crosses the y-axis at point $(0, b)$ with a slope of m, this information can be substituted into the point-slope equation and rearranged.

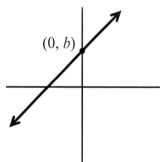

$$y - y_1 = m(x - x_1)$$

$$y - b = m(x - 0)$$

$$y - b = m(x)$$
$$\underline{+b \qquad +b}$$

$$y = mx + b$$

This is called the **slope-intercept equation** of a line, where m is the slope of the line, and b is the y-intercept of line.

The answer in point-slope form from **example 1** can be rearranged into slope-intercept form as follows:

$$y - 3 = \frac{2}{3}(x - 4)$$

$$y - 3 = \frac{2}{3}(x) - \frac{2}{3}(4)$$

$$y - 3 = \frac{2}{3}x - \frac{8}{3}$$
$$\phantom{y - 3 = \frac{2}{3}x}+3 \qquad +3$$

$$y = \frac{2}{3}x - \frac{8}{3} + \frac{9}{3}$$

$$y = \frac{2}{3}x + \frac{1}{3}$$

A line that passes through point (4, 3) and has a slope of $\frac{2}{3}$ crosses the y-axis at point $\left(0, \frac{1}{3}\right)$.

---

example 2

What is the y-intercept of the graph of the line represented by the equation below?

$$y = \frac{4}{5}x - 2$$

The easiest way to determine the slope or y-intercept of a line based on its equation is to re-write the equation in slope-intercept form, or in other words, solve the equation for y.

The equation in this problem is already in slope-intercept form:

$$y = mx + b$$

where $m = \frac{4}{5}$ and $b = -2$.

Remember that m represents the slope of the line, and b represents the y-value of the y-intercept.

The y-intercept is (0, −2).

THIS PAGE INTENTIONALLY BLANK

Practice Problems

Name _____

Write the point-slope form of the equation of a line that has the given slope and passes through the given point.

1) $(4, 2)$, $m = 5$

2) $(0, 0)$, $m = -\dfrac{2}{3}$

3) $(0, -5)$, $m = 3$

4) $(-3, -1)$, $m = \dfrac{4}{5}$

Write the point-slope form of the equation of a line that passes through the given points.

5) $(0, -5)$, $(2, -4)$

6) $(5, 1)$, $(2, 3)$

7) $(3, 2)$, $(1, 0)$

8) $(0, 4)$, $(-2, 6)$

Name the slope and y-intercept of the line represented by the given equation.

9) $y = \dfrac{2}{3}x - 4$

10) $y = -\dfrac{1}{2}x + 5$

Write the following equations of a line in standard form.

11) $y = \dfrac{2}{3}x - 6$

12) $y = -3x + \dfrac{1}{4}$

Write the following equations of a line in slope-intercept form.

13) $2x - 4y = 3$

14) $y - 4x = -2$

Write the slope-intercept form of the equation of a line that has the given slope and passes through the given point.

15) $(-1, -3), m = \dfrac{1}{3}$

16) $(2, 0), m = -\dfrac{5}{2}$

Write the equation of a line in standard form that has the given slope and passes through the given point.

17) $(-2, 1), m = 3$

18) $(-4, -1), m = -\dfrac{2}{3}$

Write the point-slope form of the equation of a line that passes through the given points.

19) $(3, -4), (0, -5)$

20) $(4, -1), (2, 3)$

Write the slope-intercept form of the equation of a line that passes through the given points.

21) $(-4, 8), (3, 5)$

22) $(0, -3), (4, -3)$

Write the equation of a line in standard form that passes through the given points.

23) $(-1, 4), (-2, 4)$

24) $(3, 3), (1, 6)$

Lesson 18: Graphing

<u>Lines</u>

To graph a linear equation, there are <u>three</u> methods from which to choose. It's good to be familiar with all three methods to be able to handle *any* graphing situation.

All three methods for graphing the equation $2x + 3y = 12$ will be demonstrated below:

▶ *Method 1*

Solve the equation in terms of y, and then create a list of ordered pairs to be graphed by substituting random values for x into the equation and solving for y.

$$2x + 3y = 12$$
$$-2x \qquad -2x$$

$$3y = -2x + 12$$

$$\frac{3y}{3} = \frac{-2x+12}{3} = \frac{-2x}{3} + \frac{12}{3}$$

$$y = -\frac{2}{3}x + 4$$

Choose values for x (small positive and negative integers are recommended) to substitute into this new equation and solve for y.

$$\text{if } x = -3 \quad \Rightarrow \quad y = -\frac{2}{3}(-3) + 4$$

$$y = 2 + 4$$

$$y = 6$$

Make a table of x- and y-values to keep these numbers organized:

x	-3	-2	-1	0	1	2	3
y	6	$\frac{16}{3}$ or $5\frac{1}{3}$	$\frac{14}{3}$ or $4\frac{2}{3}$	4	$\frac{10}{3}$ or $3\frac{1}{3}$	$\frac{8}{3}$ or $2\frac{2}{3}$	2

Some of the resulting y-values are fractions in this example, and you don't need *seven* points to graph a line. Only two points are really needed (although finding three or four is recommended in case any errors are made).

Also, try to choose x-values in a way that avoids getting fractions for y-values since fractions are difficult to plot on a graph.

Now plot the integer ordered pairs and connect the dots:

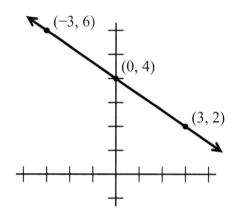

This method of plotting ordered pairs first can take longer than the other methods to follow, but this method can be applied to *any kind of equation*, not just linear equations, so it is highly recommended that you can graph a line in this way.

▶ *Method 2*

Solve the equation in terms of *y*, and with the equation now in slope-intercept form ($y = mx + b$), plot the *y*-intercept point (0, *b*) and use the slope ($m = rise$ over *run*) to find a second point to plot. Then connect the dots to graph the line.

$$2x + 3y = 12$$
$$\underline{-2x \qquad\quad -2x}$$

$$3y = -2x + 12$$

$$\frac{3y}{3} = \frac{-2x + 12}{3}$$

$$y = -\frac{2}{3}x + 4$$

From this equation, we can see that the *y*-intercept is (0, 4) and the slope of the line is $-\frac{2}{3}$.

Now plot the *y*-intercept:

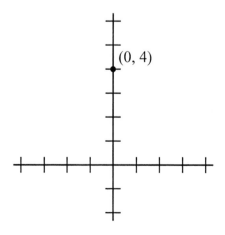

Then use the *slope* to find a second point. This can be done in two ways:

$$-\frac{2}{3} \text{ can be written as } \frac{-2}{3} \text{ or } \frac{2}{-3}$$

1) $\dfrac{-2}{3}$ means that the rise is -2 while the run is 3

2) $\dfrac{2}{-3}$ means that the rise is 2 while the run is -3

Plot a second point by going down 2 and to the right 3 from point (0, 4), or by going up 2 and to the left 3 from the *y*-intercept. Next, connect the points to graph the line:

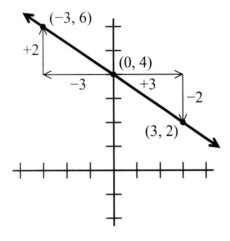

▶ *Method 3*

Find the *x*- and *y*-intercepts of the line, plot those two points, and then graph the line by connecting the points.

If the equation you are graphing is given in *standard form*, no rearranging of the equation is needed. In this example, the equation *was* given in standard form: $2x + 3y = 12$

The *x-intercept* is where a line crosses the *x*-axis and has the form (*x*, 0). The *y*-value is zero because $y = 0$ everywhere on the *x*-axis.

To calculate the *x*-value of the *x*-intercept, set $y = 0$ in the equation you are graphing and solve for *x*:

$$2x + 3y = 12$$

$$2x + 3(0) = 12$$

$$2x = 12$$

$$\frac{2x}{2} = \frac{12}{2}$$

$$x = 6 \qquad \Rightarrow \qquad \text{The } x\text{-intercept is } (6, 0).$$

The *y-intercept* is where a line crosses the *y*-axis and has the form $(0, y)$. The *x*-value is zero because $x = 0$ everywhere on the *y*-axis.

To calculate the *y*-value of the *y*-intercept, set $x = 0$ in the equation you are graphing and solve for *y*:

$$2x + 3y = 12$$

$$2(0) + 3y = 12$$

$$3y = 12$$

$$\frac{3y}{3} = \frac{12}{3}$$

$$y = 4 \qquad \Rightarrow \qquad \text{The } y\text{-intercept is } (0, 4).$$

Plot these two points, $(6, 0)$ and $(0, 4)$, and connect the dots to draw the line.

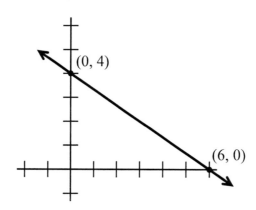

Methods 2 and 3 are the easiest to use, and which one you choose will depend on what form the equation is in when it is given, or what form is it *almost* in (slope-intercept form or standard form).

If the equation is already in slope-intercept form, or can be easily rearranged into slope-intercept form, use method 2.

If the equation is already in standard form, or close to standard form, use method 3.

Parabolas

While a line is the graph of an equation with one or two variables, each raised to the power of 1, (such as $y = 2x + 3$, or $y^1 = 2x^1 + 3$, although 1's are usually never written as exponents), a **parabola** is the graph of an equation where one of the variables is *squared*, or in other words, raised to the power of 2 (such as $y = x^2$).

A graph is a way of showing visually how one variable compares to another in an equation. When values are substituted into an equation for x, the resulting expression can be simplified to see what the corresponding value of y is. These matching values are called an "ordered pair", and they not only match together x- and y-values that are related to each other through an equation, they indicate the coordinates of a point on a graph. (They are called an *ordered* pair because the order of the numbers is important. (3, 4) means $x = 3$ and $y = 4$, but (4, 3) means $x = 4$ and $y = 3$.)

An equation like $y = 2x + 3$ is called a linear equation because graphing it will produce a line. An equation like $y = x^2$ is called a **quadratic equation** and graphing it will produce a parabola. (Note: The prefix "quad" usually refers to *four* of something, but because a variable is being *squared*, the equation is called "quadratic" from the Latin word *quadrus*, which means "square" — a shape with four sides.)

If we create a list of ordered pairs from the equation $y = x^2$, we can make a list of points that can be plotted on a graph:

x	y
–3	9
–2	4
–1	1
0	0
1	1
2	4
3	9

Plotting these points gives us the following graph:

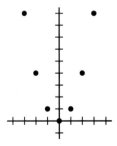

If we then connect the dots, the curved shape that results is a parabola:

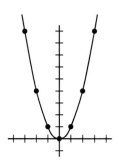

The graph of *every* quadratic equation will be a parabola. The **vertex** of the parabola (which is the "bottom" of the parabola) may be somewhere else on the graph besides point (0, 0), or the parabola may be upside down, but if it's the graph of an equation with x^2 in it (but not x^3 or higher), then it will always be a parabola.

To graph any other quadratic equation (besides something simple like $y = x^2$), just choose a few values of x to plug into the equation and find the matching value for y. Then plot these ordered pairs on a graph and connect the dots, but just remember that the graph will be a *curve*.

example 1

Graph the equation $y = x^2 - 4x + 3$

First, come up with list of ordered pairs by picking a few values for x and using them in the equation to solve for y:

$$x = -1 \qquad y = (-1)^2 - 4(-1) + 3 \qquad y = 1 + 4 + 3 = 8$$
$$x = 0 \qquad y = (0)^2 - 4(0) + 3 \qquad y = 0 - 0 + 3 = 3$$
$$x = 1 \qquad y = (1)^2 - 4(1) + 3 \qquad y = 1 - 4 + 3 = 0$$
$$x = 2 \qquad y = (2)^2 - 4(2) + 3 \qquad y = 4 - 8 + 3 = -1$$
$$x = 3 \qquad y = (3)^2 - 4(3) + 3 \qquad y = 9 - 12 + 3 = 0$$
$$x = 4 \qquad y = (4)^2 - 4(4) + 3 \qquad y = 16 - 16 + 3 = 3$$
$$x = 5 \qquad y = (5)^2 - 4(5) + 3 \qquad y = 25 - 20 + 3 = 8$$

Now plot these ordered pairs and trace a curve through them:

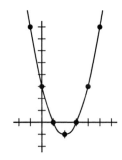

Practice Problems

Name _____

Graph each equation by plotting at least three ordered pairs.

1) $y = 2x - 1$

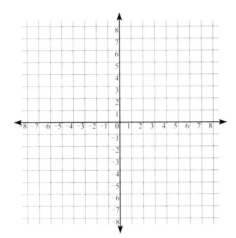

2) $x + y = 3$

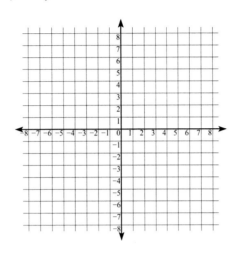

Graph each equation by using the x-intercept and y-intercept.

3) $x - y = 4$

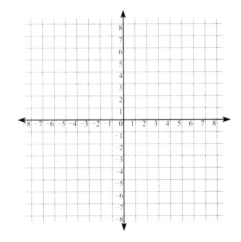

4) $3x + y = 6$

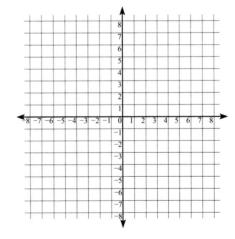

Graph each equation by using the slope and *y*-intercept.

5) $y = -\dfrac{5}{2}x + 1$

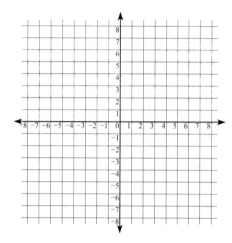

6) $y = \dfrac{2}{3}x - 2$

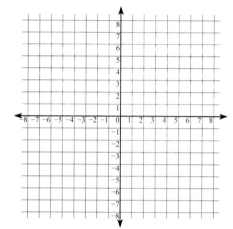

Graph each quadratic equation.

7) $y = x^2 - 2x - 3$

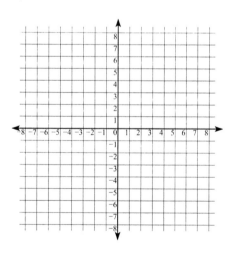

8) $y = -x^2 + 3$

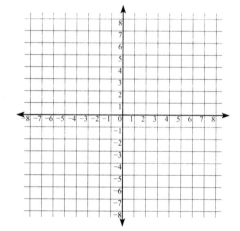

Lesson 19: Parallel and Perpendicular Lines

Parallel Lines

Parallel lines have the *same slope* but different *y*-intercepts.

The following two equations are parallel lines when graphed:

$$y = 2x + 3; \ y = 2x - 1$$

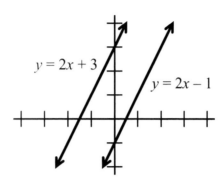

Checking if two lines are parallel

If two linear equations are written in slope-intercept form, compare the slopes, and if they are the same then the lines are parallel. However, if two linear equations are not written in slope-intercept form, rearrange both equations into slope-intercept form and compare the slopes.

---

example 1

Does $3x + 2y = 4$ represent a line that is parallel to $3x + 4y = 7$?

Rearrange each equation into slope-intercept form and compare the slopes:

$$3x + 2y = 4 \qquad\qquad\qquad 3x + 2y = 7$$
$$-3x \qquad -3x \qquad\qquad\quad -3x \qquad\quad -3x$$

$$2y = -3x + 4 \qquad\qquad\quad 2y = -3x + 7$$

$$\frac{2y}{2} = \frac{-3x + 4}{2} \qquad\qquad \frac{2y}{2} = \frac{-3x + 7}{2}$$

$$y = -\frac{3}{2}x + 2 \qquad\qquad\quad y = -\frac{3}{2}x + \frac{7}{2}$$

The slopes of both lines are $-\dfrac{3}{2}$, so the equations *do* represent parallel lines.

Perpendicular Lines

Perpendicular lines have slopes that are negative reciprocals of each other.

The following two equations are perpendicular lines when graphed:

$$y = 3x + 4; \; y = -\frac{1}{3}x + 2$$

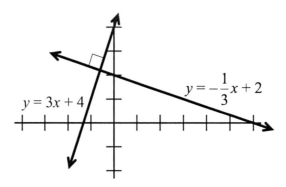

Checking if two lines are perpendicular

If two linear equations are written in slope-intercept form, compare the slopes, and if they are negative reciprocals of each other then the lines are perpendicular. However, if two linear equations are not written in slope-intercept form, rearrange both equations into slope-intercept form and compare the slopes.

---

example 2

Does $2x - 3y = 1$ represent a line that is perpendicular to $3x + 2y = 4$?

Rearrange each equation into slope-intercept form and compare the slopes:

$$2x - 3y = 1 \qquad\qquad\qquad 3x + 2y = 4$$
$$-2x \qquad -2x \qquad\qquad\quad -3x \qquad\quad -3x$$

$$-3y = -2x + 1 \qquad\qquad\quad 2y = -3x + 4$$

$$\frac{-3y}{-3} = \frac{-2x+1}{-3} \qquad\qquad \frac{2y}{2} = \frac{-3x+4}{2}$$

$$y = \frac{2}{3}x - \frac{1}{3} \qquad\qquad\quad y = -\frac{3}{2}x + 2$$

The slope of the first line is $\frac{2}{3}$, and the slope of the second line is $-\frac{3}{2}$.

Since $\frac{2}{3}$ is the negative reciprocal of $-\frac{3}{2}$, the equations *do* represent perpendicular lines.

Finding the Equation of a Line Parallel or Perpendicular to a Given Line

---

example 3

Write an equation in slope-intercept form of the line that passes through point $(2, -3)$ and is parallel to the graph of $4x - y = 6$.

First find the slope of the line of the given equation:

$$4x - y = 6$$
$$-4x \qquad -4x$$

$$-y = -4x + 6$$

$$(-1)(-y) = (-4x + 6)(-1)$$

$$y = 4x - 6 \qquad\qquad \Rightarrow \qquad$$ The slope of the given line is 4, so the slope of the parallel line will also be 4.

Next, substitute the given point and slope into the point-slope equation of a line:

$$y - y_1 = m(x - x_1)$$

$$y - (-3) = 4(x - 2)$$

$$y + 3 = 4x - 8$$
$$ -3 \qquad\quad -3$$

$$y = 4x - 11$$

If you were asked to find an equation of a line that passes through a given point but is *perpendicular* to a given equation, follow the same steps as above, but when you find the slope of the line of the given equation, remember to use the *negative reciprocal* of that slope to find the answer.

THIS PAGE INTENTIONALLY BLANK

Practice Problems

Name _____

Write an equation (in slope-intercept form) for the line that passes through the given point and is parallel to the graph of the given equation.

1) $(2, -1)$; $y = 2x - 1$

2) $(-4, 0)$; $y = -\dfrac{1}{3}x + 4$

3) $(-7, 2)$; $3x + y = 7$

4) $(1, 8)$; $-4x - y = 5$

5) $(-1, 4)$; $y = 2$

6) $(3, 4)$; $5x - 4y = 2$

7) $(-3, -2)$; $2x - 7y = -3$

8) $(6, 2)$; $x = -5$

Write an equation (in slope-intercept form) for the line that passes through the given point and is perpendicular to the graph of the given equation.

9) $(0, -1)$; $y = \dfrac{2}{5}x - 3$

10) $(0, 0)$; $x = 3$

11) $(1, 4)$; $y = -4x - 7$

12) $(3, -3)$; $5x + y = -9$

13) $(-2, 5)$; $\dfrac{3}{2}x + y = 1$

14) $(-4, 2)$; $x - 5y = 2$

15) $(0, 3)$; $y = \dfrac{1}{2}$

16) $(-3, 4)$; $7x + 4y = -5$

Lesson 20: Solving Equations

To solve an equation, you must follow certain rules, and there are several guidelines that are suggested you follow.

The rules for solving an equation (of one variable) can be summed up in one statement. You can do anything you want to an equation, but _what you do to one side of the equation, you must to do the other side of the equation_.

This can be seen in the following four properties or "rules":

1) Addition Property of Equality

   For any numbers a, b, and c, if $a = b$, then $a + c = b + c$.

2) Subtraction Property of Equality

   For any numbers a, b, and c, if $a = b$, then $a - c = b - c$.

3) Multiplication Property of Equality

   For any numbers a, b, and c $(c \neq 0)$, if $a = b$, then $ac = bc$.

4) Division Property of Equality

   For any numbers a, b, and c $(c \neq 0)$, if $a = b$, then $\dfrac{a}{c} = \dfrac{b}{c}$.

When using the multiplication or division properties of equality, remember that if there is more than one term on either side of the equation, _every term_ on each side of the equation must be multiplied or divided by whatever c is.

The following guidelines are suggested for solving an equation:

1) If any distributing is required, do that first.

2) Combine all **like terms** on either side of the equation (and _always_ combine any like terms before moving terms from one side of the equation to the other).

3) Move all terms containing a variable to one side of the equation (usually the left side).

4) Move all **constants** (terms without variables) to the other side of the equation (usually the right side).

5) Multiply or divide to isolate the variable on one side of the equation.

---

example 1

Solve for x: $2(x - 3) + 4x = 3x - 7$

1) Distribute: $2x - 6 + 4x = 3x - 7$

2) Combine like terms: $6x - 6 = 3x - 7$

3) Move all variable terms $6x - 6 = 3x - 7$
   to one side: $-3x -3x$

                                  $3x - 6 = -7$

4) Move all constants to $3x - 6 = -7$
   the other side: $+6 +6$

                                  $3x = -1$

5) Divide by the coefficient
   of x to isolate the variable: $\dfrac{3x}{3} = \dfrac{-1}{3}$

                                  $x = -\dfrac{1}{3}$

Common Errors When Solving Multi-step Equations

If you first undo any multiplications or divisions involving the variable, make sure you multiply or divide *every term*, and not just the term with the variable. This is a common error in algebra:

---

example 2

What does x equal in this equation?

$$\frac{x}{4} + 8 = 32$$

There are two ways to start this problem: 1) Subtract 8 from both sides, or 2) Multiply both sides by 4.

Method 1

$$\frac{x}{4} + 8 = 32$$
$$\phantom{\frac{x}{4}}-8 -8$$

$$\frac{x}{4} = 24$$

If you undo any additions or subtractions first, you can then undo any multiplications or divisions involving the variable.

$$4 \cdot \frac{x}{4} = 24 \cdot 4$$

$$x = 96$$

Method 2

$$\frac{x}{4} + 8 = 32$$

If you multiply both sides by 4 first:

<table>
<tr><td align="center">right</td><td align="center">wrong</td></tr>
<tr><td>$4 \cdot \left(\frac{x}{4} + 8\right) = 32 \cdot 4$</td><td>$4 \cdot \left(\frac{x}{4} + 8\right) = 32 \cdot 4$</td></tr>
<tr><td>$4\left(\frac{x}{4}\right) + 4(8) = 128$</td><td>$x + 8 = 128$</td></tr>
<tr><td>$x + 32 = 128$
$-32 \quad -32$</td><td>$x + 8 = 128$
$-8 \quad -8$</td></tr>
<tr><td align="center">$x = 96$</td><td align="center">$x = 120$</td></tr>
</table>

A common error in the problem below is to mistakenly distribute to the second term of the numerator in the fraction on the left side of the equation:

example 3a

Solve for x: $\quad \dfrac{3x + 2}{4} = 2$

There are two ways to start this problem:

1) Multiply both sides by 4, or

2) Subtract $\dfrac{1}{2}$ from both sides

Method 1

$$\frac{3x + 2}{4} = 2$$

If you multiply both sides by 4 first:

<u>right</u>   <u>wrong</u>

$$4 \cdot \left(\frac{3x+2}{4}\right) = 2 \cdot 4 \qquad\qquad 4 \cdot \left(\frac{3x+2}{4}\right) = 2 \cdot 4$$

$$\begin{array}{ll} 3x + 2 = 8 & \qquad\qquad 3x + 8 = 8 \\ \underline{-2\ -2} & \qquad\qquad \underline{-8\ -8} \end{array}$$

$$3x = 6 \qquad\qquad\qquad 3x = 0$$

$$\frac{3x}{3} = \frac{6}{3} \qquad\qquad\qquad \frac{3x}{3} = \frac{0}{3}$$

$$x = 2 \qquad\qquad\qquad x = 0$$

Another common mistake is to subtract a term from a numerator while ignoring the fact that *each term of the numerator* is being divided by the denominator:

example 3b

Method 2

The original problem can be re-written as:

$$\frac{3x}{4} + \frac{2}{4} = 2$$

If you subtract from both sides first:

<u>right</u>   <u>wrong</u>

$$\frac{3x}{4} + \frac{1}{2} = 2 \qquad\qquad \frac{3x+2}{4} = 2$$

$$\begin{array}{ll} \underline{-\dfrac{1}{2} \ -\dfrac{1}{2}} & \qquad\qquad \underline{-2 \ -2} \end{array}$$

$$\frac{3x}{4} = \frac{3}{2} \qquad\qquad\qquad \frac{3x}{4} = 0$$

$$4 \cdot \left(\frac{3x}{4}\right) = \left(\frac{3}{2}\right) \cdot 4 \qquad\qquad 4 \cdot \left(\frac{3x}{4}\right) = 0 \cdot 4$$

$$3x = 6 \qquad\qquad\qquad 3x = 0$$

$$\frac{3x}{3} = \frac{6}{3} \qquad\qquad\qquad \frac{3x}{3} = \frac{0}{3}$$

$$x = 2 \qquad\qquad\qquad x = 0$$

Practice Problems

Name _____

Solve each equation. Remember to check each answer.

1) $b + 8 = 22$

2) $x - 14 = -14$

3) $d - 8 = 14$

4) $\dfrac{4}{9} = a + \dfrac{1}{9}$

5) $22 = 32 - f$

6) $m - \dfrac{3}{5} = \dfrac{2}{3}$

7) $5n = 40$

8) $3x = 18$

9) $\dfrac{b}{5} = 7$

10) $-2d = 14$

11) $-\dfrac{h}{4} = 4$

12) $5g = -\dfrac{2}{3}$

13) $5x + 14 = 49$

14) $10m - 9 = 31$

15) $15 - 6c = -33$

Solve each equation. Remember to check each answer.

16) $4g - 7 = 21$

17) $5 = 3d - 10$

18) $6 - 5x = -19$

19) $\dfrac{x}{3} - 6 = -2$

20) $-\dfrac{a}{4} + 7 = 7$

21) $-12 = 6 + \dfrac{2}{3}m$

22) $-8 = \dfrac{t + 12}{-3}$

23) $\dfrac{6a + 8}{5} = 10$

24) $-3 = -\dfrac{p}{4} - 9$

Write an equation and solve. Remember to check each answer.

25) Find three consecutive odd integers whose sum is 87. Find the integers.

26) Twice the greater of two consecutive even integers is 8 less than four times the lesser integer. Find the integers.

27) The perimeter of a rectangle is 36 inches. Find the dimensions if its length is 4 inches longer than its width.

Lesson 21: Solving Inequalities

All of the same equality properties of addition and subtraction that apply to equations also apply to inequalities.

For any numbers a, b, and c:

1) If $a > b$, then $a + c > b + c$ <u>or</u> if $a < b$, then $a + b < b + c$.

2) If $a > b$, then $a - c > b - c$ <u>or</u> if $a < b$, then $a - b < b - c$.

The *one* difference between equations and inequalities is with the multiplication and division properties of equality.

<u>Multiplication Property for Inequalities</u>

For any numbers a, b, and c:

<u>when c is positive</u>: if $a > b$, then $ac > bc$ <u>or</u> if $a < b$, then $ac < bc$.

<u>when c is negative</u>: if $a > b$, then $ac < bc$ <u>or</u> if $a < b$, then $ac > bc$.

When multiplying both sides of an inequality by a *negative number*, the inequality symbol flips around ("<" becomes ">", or ">" becomes "<").

---

example 1

Solve for x:

$$6 - \frac{x}{2} > 1$$
$$-6 \qquad -6$$

$$-\frac{x}{2} > -5$$

$$(-2)\left(-\frac{x}{2}\right) > (-5)(-2)$$

$\downarrow$ at this step the inequality symbol changes because *both sides* were multiplied by a negative number

$$x < 10$$

This also happens if both sides of an inequality are *divided* by a negative number.

Division Property for Inequalities

For any numbers a, b, and c:

when c is positive: if $a > b$, then $\dfrac{a}{c} > \dfrac{b}{c}$ _or_ if $a < b$, then $\dfrac{a}{c} < \dfrac{b}{c}$.

when c is negative: if $a > b$, then $\dfrac{a}{c} < \dfrac{b}{c}$ _or_ if $a < b$, then $\dfrac{a}{c} > \dfrac{b}{c}$.

example 2

Solve for x:

$$-3x + 1 \geq 7$$
$$ -1 \ -1$$

$$-3x \geq 6$$

$$\frac{-3x}{-3} \geq \frac{6}{-3}$$

$\downarrow$ at this step the inequality symbol changes because _both sides_ were divided by a negative number

$$x \leq -2$$

Graphing an Inequality on a Number Line

example 3

Which graph below represents the solution to the inequality below?

$$2(2x - 6) \geq x + 3$$

A.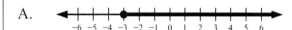

B.

C.

D.

First, solve the inequality for x by distributing the 2 on the left side, moving all terms with the variable to the left side and all the constants to the right side, and then isolate the variable:

$$2(2x - 6) \geq x + 3$$

$$4x - 12 \geq x + 3$$
$$\underline{-x} \qquad \underline{-x}$$

$$3x - 12 \geq 3$$
$$\underline{+12 \ +12}$$

$$3x \geq 15$$

$$\frac{3x}{3} \geq \frac{15}{3}$$

$$x \geq 5$$

Next, determine which number line represents this solution. Because x is greater than *or equal to* 5, a small *closed* circle will be drawn at the 5 on the number line, and the line will extend to the right where numbers get larger.

The answer is D.

example 4

Imelda will work 10 to 20 hours per week at her new job and will be paid $7.50 per hour. Which of the following shows how much she can earn per week?

A.

B.

C.

D.

If Imelda works only 10 hours, she will make $75 (10 × $7.50 = $75). If she works 20 hours in a week, she will make $150 (20 × $7.50 = $150). By using the information in the problem to find out what the minimum amount of money and maximum amount of money she could make, we get numbers we can use in the following inequality:

$$\$75 \leq x \leq \$150, \text{ where } x \text{ is the amount of money Imelda can make.}$$

Of the four answers given, only two (A and C) show a solution with values *between* $75 and $150. Open circles over $75 and $150 mean that Imelda could not make exactly either of those amounts, while closed circles mean that she could make exactly $75 or $150. Since it is possible for her to make either of those exact amounts (if she works 10 hours or 20 hours), the circles should be closed.

The answer is C.

Practice Problems

Name _____

Solve each inequality. Remember to check each answer. Then graph each answer on a number line.

1) $b + 4 < 13$

2) $10 \leq 5 - c$

3) $-12 + a \geq 2a - 5$

4) $\dfrac{y}{4} \geq 2$

5) $-\dfrac{n}{6} < -5$

6) $2p < -11$

7) $\dfrac{3d}{5} - 2 < 4$

8) $\dfrac{3t - 3}{4} \geq -6$

Solve each inequality. Remember to check each answer. Then graph each answer on a number line.

9) $4n - 6 < 6n - 10$

10) $\dfrac{z+3}{2} < -6$

11) $6s + 6 > 4 - (s + 12)$

12) $3b + 11 < -(2b - 6)$

Write an inequality and solve. Remember to check each answer.

13) Three times a number is no more than 48.

14) Six more than the quotient of a number and 2 is at least that number.

15) The sum of a number and 24 is greater than the product of −3 and that number.

Lesson 22: Solving Equations and Inequalities with Absolute Value

To solve an equation or inequality with a variable within an absolute value, *always isolate the absolute value on one side of the equation or inequality first.*

Equations Involving Absolute Value

An equation with a variable within an absolute value can be rewritten as *two* equations.

---

example 1

Solve for x:

$$|x+2| = 5$$

The absolute value symbol means that the positive *or* negative form of $(x + 2)$ equals 5.

The original equation can be rewritten as:

$x + 2 = 5$ \qquad or \qquad $-(x + 2) = 5$

⇑

If both sides are multiplied by -1, this can be rewritten as:

$x + 2 = -5$

Solving each of these will provide two solutions to the original equation.

$$
\begin{array}{lcl}
x + 2 = 5 & & x + 2 = -5 \\
\underline{-2 \ -2} & \text{or} & \underline{-2 \ \ -2} \\
x = 3 & \text{or} & x = -7
\end{array}
$$

Note: There is no solution to $|x+2| = -5$.

Inequalities Involving Absolute Value

An inequality with a variable within an absolute value can be rewritten as a **compound inequality**.

There are **two** kinds of compound inequalities that can be created.

1) If the inequality has a "$<$" or "$\leq$", then once the absolute value is isolated (on the left-hand side), the compound inequality will have an **AND**.

example 2

Solve for x:

$$|x-3| < 8$$

This can be rewritten as:

$$x - 3 < 8 \qquad \text{and} \qquad -(x - 3) < 8$$
$$+3 \ +3 \qquad\qquad\qquad\qquad \Uparrow$$

or rewrite this as:

$$x - 3 > -8$$
$$+3 \quad +3$$

$$x < 11 \qquad \text{and} \qquad x > -5$$

This means that all solutions must meet both criteria, which is all values less than 11 *and* greater than −5.

This can be expressed as a graph on a number line:

Note: If there was a "≤" or "≥", the circles on the number line would be closed.

2) If the inequality has a ">" or "≥", then once the absolute value is isolated (on the left-hand side), the compound inequality will have an **OR**.

example 3

Solve for x:

$$|x+2| \geq 4$$

This can be rewritten as:

$$x + 2 \geq 4 \qquad \text{or} \qquad -(x + 2) \geq 4$$
$$-2 \ -2 \qquad\qquad\qquad\qquad \Uparrow$$

or rewrite this as:

$$x + 2 \leq -4$$
$$-2 \quad -2$$

$$x \geq 2 \qquad \text{or} \qquad x \leq -6$$

This means that the solution includes all values greater than or equal to 2 *or* all values less than or equal to −6. The solution only has to meet *one* of these criteria.

This can be expressed as a graph on a number line:

Practice Problems

Name _____

Solve each equation. Remember to check each answer.

1) $|x - 7| = 4$

2) $|4 - c| = 6$

3) $|3x - 6| = 5$

4) $|2 - 4b| = 10$

5) $\left|2m + \dfrac{2}{3}\right| = -\dfrac{4}{3}$

6) $\left|3p - \dfrac{1}{4}\right| = \dfrac{5}{2}$

7) $|a - 3| + 5 = 8$

8) $6 + |2 + x| = 16$

9) $\left|3t - \dfrac{4}{3}\right| - 6 = \dfrac{7}{3}$

10) $2\left|x - \dfrac{2}{7}\right| = 8$

11) $\dfrac{1}{2}\left|\dfrac{3g}{4} - \dfrac{2}{5}\right| = 6$

12) $3\left|\dfrac{4k + 2}{3}\right| = \dfrac{k}{2}$

Solve each inequality. Graph each solution set on a number line.

13) $|x+3| > 5$

14) $|2-m| < 3$

15) $|4x| + 2 \geq 26$

16) $2|x-5| \leq 8$

17) $|3d-9| < 3$

18) $|4+z| \geq 1$

19) $\dfrac{|2f+3|}{4} > 3$

20) $\left|\dfrac{4}{3} - 2c\right| \leq \dfrac{3}{2}$

21) $\left|\dfrac{5-2x}{3}\right| \geq -\dfrac{5}{3}$

22) $\left|\dfrac{5f}{3}\right| - \dfrac{1}{4} < \dfrac{7}{2}$

Lesson 23: Formulas

A **formula** is an equation that contains more than one variable and defines the relationship among those variables. A formula can be solved in terms of any of the variables in that formula.

For example, the distance a vehicle travels is determined by multiplying its rate of travel (its speed) by the time it is traveling:

$$d = r \times t$$

(distance = rate × time)

This formula has three unknowns (variables), so as long as information is provided about two of the variables, the formula can be used to solve for the third variable.

example 1

As given, the above formula is solved for *d*, distance. To solve the formula in terms of rate, divide both sides by *t*, time.

$$d = r \cdot t$$

$$\frac{d}{t} = \frac{r \cdot t}{t}$$

$$r = \frac{d}{t}$$

example 2

Or to solve the formula in terms of *t*, time, divide both sides of the original formula by *r*, the rate.

$$d = r \cdot t$$

$$\frac{d}{r} = \frac{r \cdot t}{r}$$

$$t = \frac{d}{r}$$

The same equality properties used to solve equations are also used to rearrange formulas to be solved in terms of a specific variable.

Understanding Formulas

---

example 3

The following equation shows the relationship between a distance traveled (d), the time traveled (t), and the rate (r):

$$\frac{d}{t} = r$$

If the time t increases and the distance d remains the same, what happens to the rate r?

A. It increases.
B. It decreases.
C. It remains the same.
D. There is not enough information given to tell.

When the numerator of a fraction is held constant, the value of the fraction *decreases* as the denominator *increases*:

$$\frac{1}{2} > \frac{1}{3} > \frac{1}{4} > \frac{1}{5} \text{ and so on…}$$

So if t is increasing in the formula, while d remains the same, r would decrease.

The answer is B.

---

example 4

In the equation shown below, x represents a positive real number.

$$y = \frac{100}{x} + 50$$

As the value of x gets larger, what happens to the value of y?

A. The value of y stays the same.
B. The value of y increases.
C. The value of y approaches 50.
D. The value of y approaches 100.

Just like the last example, this formula has a variable in the denominator of a fraction and the question is asking about what happens when the value of that variable increases.

When x has a value of less than 100 (but greater than zero, and remember, a denominator can *never* equal zero), the fraction $\dfrac{100}{x}$ has a value greater than 1. For example:

$$\frac{100}{1} = 100, \ \frac{100}{5} = 20, \ \frac{100}{10} = 10, \ \frac{100}{20} = 5, \ \frac{100}{50} = 2, \ \frac{100}{100} = 1$$

Notice how the value of the fraction *decreases* as the denominator *increases*. Also notice how the value of the fraction will become less than 1 if the denominator is greater than 100:

$$\frac{100}{100} = 1, \ \frac{100}{200} = 0.5, \ \frac{100}{1000} = 0.1, \ \frac{100}{10,000} = 0.01, \ \frac{100}{100,000} = 0.001$$

As x (the denominator) gets larger and larger, the overall value of the fraction gets closer and closer to zero. If this term in the formula is getting smaller as x gets larger, a smaller and smaller value is being added to 50. So the value of y will get closer to 50 as x decreases.

The answer is C.

---

example 5

The formula for the volume (V) of a cube is

$$V = e^3$$

where e is the length of an edge.

An edge of a silver cube is twice as long as an edge of a gold cube. How many times greater is the volume of a silver cube than that of a gold cube?

A. 2 times greater
B. 9 times greater
C. 8 times greater
D. 6 times greater

Instead of thinking about how the edge of the silver cube is double that of the gold cube, let's say that the edge of a gold cube has a measure of 1 unit and the edge of a silver cube has a measure of 2 units. With an edge with a length of 1 unit, the gold cube would have a volume of 1 cubic unit. ($V = 1^3 \Rightarrow V = 1$) With an edge with a length of 2 units, the volume of the silver cube would be 8 cubic units. ($V = 2^3 \Rightarrow V = 8$)

The answer is C.

example 6

Solve the following formula in terms of x.

$$\frac{ax - by}{c} = z$$

$$c \cdot \left(\frac{ax - by}{c}\right) = z \cdot c$$

$$ax - by = cz$$
$$ +by +by$$

$$ax = by + cz$$

$$\frac{ax}{a} = \frac{by + cz}{a}$$

$$x = \frac{by + cz}{a}$$

<u>Using Formulas</u>

example 7

An object is dropped from a small plane flying at a height of 1000 feet above the ground. As the object falls, d, its distance above the ground after t seconds, is given by the formula below.

$$d = -16t^2 + 1000$$

How far above the ground is the object when it has fallen for 4 seconds?

A. 984 feet
B. 936 feet
C. 872 feet
D. 744 feet

To answer this question, you just need to substitute the 4 in for t and simplify the expression, or in other words, solve for d:

$$d = -16(4)^2 + 1000$$
$$d = -16(16) + 1000$$
$$d = -256 + 1000$$
$$d = 744$$

The answer is D.

Practice Problems

Name _____

Solve for x.

1) $2x - 5 = y$

2) $3 - ax = h$

3) $bx - f = m$

4) $\dfrac{12 - x}{a} = -9$

5) $\dfrac{4x - 5}{2} = b$

6) $\dfrac{c}{4} + \dfrac{x}{2} = \dfrac{d}{3}$

7) $\dfrac{h - 2x}{y} = 4d$

8) $\dfrac{x - y}{m} = -n$

9) $x(a + 4) = b$

10) Solve for m: $F = ma$

11) Solve for n: $pV = nRT$

12) Solve for f: $T = \dfrac{1}{f}$

13) Solve for v: $K = \dfrac{1}{2}mv^2$

14) Use the given formula to solve. What is the value of v when $v_0 = 6$, $a = 10$, and $t = 3$?

$$v = v_0 + at$$

Lesson 24: Multiplying and Dividing Monomials

Multiplying Monomials

A **monomial** is a term made from one or more numbers and/or variables multiplied together. Any variables in a monomial must have a *positive integer exponent*. (Variables can have negative exponents, but such terms wouldn't be considered to be monomials.)

examples of monomials
$7, \ 2x^2, \ 6abc^4$

When monomials are multiplied together, the following rules must be used:

Product of Powers

For any variable x, and all integers m and n:

$$x^m \cdot x^n = x^{m+n}$$

example 1
$d^2 \cdot d^4 = d^6$

You can check this answer by writing out all of the variables:

$d^2 = d \cdot d$
$d^4 = d \cdot d \cdot d \cdot d$
$(d \cdot d) \times (d \cdot d \cdot d \cdot d) = d \cdot d \cdot d \cdot d \cdot d \cdot d$ or d^6

Power of a Power

For any variable x, and all integers m and n:

$$(x^m)^n = x^{m \cdot n}$$

example 2
$(a^4)^3 = a^4 \cdot a^4 \cdot a^4 = a^{12}$

Power of a Product

For variables x and y, and any integer m:

$$(xy)^m = x^m y^m$$

example 3

$$(ab)^2 = (ab)(ab) = (a \cdot a)(b \cdot b) = a^2 b^2$$

example 4

$$(2x)^4 = 2^4 x^4 = 16x^4$$

Power of a Monomial

For variables x and y, and all integers m, n, and p:

$$(x^m y^n)^p = x^{mp} y^{np}$$

example 5

$$
\begin{aligned}
(x^2 y^3)^4 &= (x^2 y^3)(x^2 y^3)(x^2 y^3)(x^2 y^3) \\
&= (x^2 \cdot x^2 \cdot x^2 \cdot x^2)(y^3 \cdot y^3 \cdot y^3 \cdot y^3) \\
&= x^8 y^{12}
\end{aligned}
$$

Dividing by Monomials

When dividing by a monomial, use the following rules of exponents:

Quotient of Powers

For any variable x, and all integers m and n:

$$\frac{x^m}{x^n} = x^{m-n}$$

example 6

$$\frac{x^7}{x^2} = \frac{x \cdot x \cdot x \cdot x \cdot x \cdot x \cdot x}{x \cdot x} = x^5$$

Zero Exponent

For any variable :

$$x^0 = 1$$

This applies to larger expressions as well. *Anything* raised to the power of zero equals 1.

examples of zero exponent
$a^0 = 1$, $3^0 = 1$, $(4x^2y^3)^0 = 1$

Negative Exponents

For any variable x, and any integer n:

$$x^{-n} = \frac{1}{x^n}$$

(Note: Remember that a monomial must have a positive integer exponent, so x^{-n}, or $\frac{1}{x^n}$, is not a monomial.)

example 7
$\dfrac{b^2}{b^5} = \dfrac{b \cdot b}{b \cdot b \cdot b \cdot b \cdot b} = \dfrac{1}{b^3}$ OR $\dfrac{b^2}{b^5} = b^{2-5} = b^{-3} = \dfrac{1}{b^3}$

Typically (with the exception of scientific notation) answers should *not* be expressed with variables that have negative exponents. Express answers with *positive* exponents.

example 8
$\dfrac{a^2}{b^{-2}} = a^2 b^2$

example 9
$\dfrac{x^{-2} y^{-1}}{z^{-4}} = \dfrac{z^4}{x^2 y}$

THIS PAGE INTENTIONALLY BLANK

Practice Problems

Name _____

Simplify each expression.

1) $a^2(a^3)$

2) $(4x^2)(2x^3)$

3) $m^2n^3(m^2n^4)$

4) $c^2(c^3)(c^4)$

5) $(3a^3)^2$

6) $(2x)^2(3y)^2$

7) $\left(\dfrac{m}{n}\right)^4$

8) $-4(-3)^3(x^4)^3$

9) $(ab)^5$

10) $(x^2y^3)(xy)^4$

11) $(abc)^2(ab)^3$

12) $(-2e^2f^3)^3\left(\dfrac{1}{2}e^4f^2\right)^2$

13) $(-3)^3(m^4n^3)^2$

14) $\left(\dfrac{a^3}{b^4}\right)^3$

15) $(4x^3y)(-2x)^5$

Simplify each expression.

16) $\dfrac{a^5}{a^3}$

17) $\dfrac{x^4 y^6}{x^3 y^2}$

18) $\dfrac{4m^3}{12mn^2}$

19) $\dfrac{b^0}{b^3}$

20) $\dfrac{(m^3 n^2)^2}{(m^3 n^2)^0}$

21) $\dfrac{x^3 y^0}{x^0 y^5}$

22) $\dfrac{-8c^{-3} d^{-1}}{24}$

23) $(2a^2 b^3)^{-1}$

24) $\dfrac{16m^{-3} n^{-2}}{4m^{-5} n^{-7}}$

25) $\left(\dfrac{3a^2 b^{-1}}{4a^4 b^{-4}}\right)^{-2}$

26) $\dfrac{(2x^2 y^3)^2}{4x^{-1} y^3}$

27) $\dfrac{18m^2 n p^4}{6mn^5 p^2}$

28) $\dfrac{(3a^2 b^4)^2}{(2ab^4)^3}$

29) $(-4)^2 (3xy^3)(2xy)^{-4}$

30) $\dfrac{(3x^0 y^0)^3}{9(xy)^{-2}}$

Lesson 25: Multiplying Polynomials

When multiplying a polynomial by a monomial, the distributive property and rules of exponents are used.

example 1

Multiply: $4x^2 \cdot (2x + 3x^3)$

$\quad 4x^2(2x) + 4x^2(3x^3)$

$\quad (4 \cdot 2)(x^2 \cdot x) + (4 \cdot 3)(x^2 \cdot x^3)$

$\quad 8x^3 + 12x^5$

When two polynomials are multiplied together, each term from the first polynomial must be multiplied by each term of the second polynomial.

example 2

Multiply: $(x + 2y)(3x - 4y)$

$\quad x(3x - 4y) + 2y(3x - 4y)$

$\quad x(3x) + x(-4y) + 2y(3x) + 2y(-4y)$

$\quad 3x^2 - 4xy + 6xy - 8y^2$

these are called *like terms* (terms with the same variables all raised to the same power)

Like terms can be combined by adding their coefficients together, but *do not change the values of any exponents*!

$\quad 3x^2 + (-4 + 6)xy - 8y^2$

$\quad 3x^2 + 2xy - 8y^2$

Another method for multiplying the polynomials in this example is called the **FOIL method**.

When multiplying two binomials (polynomials with two terms), the FOIL method works as follows:

1) multiply the FIRST terms of each binomial

2) multiply the OUTER terms of the binomials

3) multiply the INNER terms of the binomials

4) multiply the LAST terms of each binomial

example 3	

Multiply: $(x + 2y)(3x - 4y)$

$$\quad\quad\;\, \text{F} \quad\quad\;\; \text{O} \quad\quad\;\; \text{I} \quad\quad\;\; \text{L}$$
$$x(3x) + x(-4y) + 2y(3x) + 2y(-4y)$$
$$3x^2 - 4xy + 6xy - 8y^2$$
$$3x^2 + 2xy - 8y^2$$

The FOIL method is useful for making sure no steps are overlooked, but it is only used when multiplying binomials.

When multiplying any other polynomials, remember to multiply *every* term of the first polynomial by *every* term of the second polynomial.

example 4	

Multiply the polynomials:

$$(4a - 3b)(2a^2 + ab - 5b^2)$$
$$4a(2a^2 + ab - 5b^2) - 3b(2a^2 + ab - 5b^2)$$
$$4a(2a^2) + 4a(ab) + 4a(-5b^2) - 3b(2a^2) - 3b(ab) - 3b(5b^2)$$
$$8a^3 + \underline{4a^2b} - \mathbf{20ab^2} \underline{- 6a^2b} - \mathbf{3ab^2} - 15b^3$$
$$8a^3 - 2a^2b - 23ab^2 + 15b^3$$

Special Products

There are some special cases involving the multiplication of two binomials. The FOIL method can be used anytime two binomials are multiplied, but understanding the following rules will make it easier to understand future lessons about factoring polynomials.

Square of a Sum

$$(x + y)^2 = (x + y)(x + y) = x^2 + 2xy + y^2$$

When a binomial is squared, the resulting polynomial is the first term of the binomial squared, twice the two terms multiplied together, and the last term of the binomial squared.

This can be proven with the FOIL method:

$$(x + y)(x + y) = x(x) + x(y) + y(x) + y(y)$$

$$= x^2 + \underline{xy} + \underline{xy} + y^2$$

the like terms to be combined are the *same*

$$= x^2 + 2xy + y^2$$

This shortcut can be used to skip a step or two in the FOIL method.

---

example 5

Multiply:

$$(2a - 3b)^2 = (2a)^2 + 2(2a)(-3b) + (-3b)^2$$

$$= 4a^2 - 12ab + 9b^2$$

Square of a Difference

$$(x - y)^2 = (x - y)(x - y) = x^2 - 2xy + y^2$$

This is the same as the **square of a sum**, except that a minus sign appears ahead of the middle term of the resulting trinomial.

---

example 6

Multiply:

$$(4c - d)^2 = (4c)^2 + 2(4c)(-d) + (-d)^2$$

$$= 16c^2 - 8cd + d^2$$

Difference of Squares

The **difference of squares** refers to the answer that results from multiplying two binomials with the same terms, but with the last term of each binomial having opposite signs:

$$(x + y)(x - y) = x^2 - y^2$$

The answer has no middle term because the like terms that result from the FOIL method cancel each other out:

$$(x + y)(x - y) = x(x) + x(-y) + y(x) + (y)(-y)$$

$$= x^2 - \underline{xy} + \underline{xy} - y^2$$

$$\uparrow \quad \uparrow$$

the like terms have opposite signs

$$= x^2 - y^2$$

THIS PAGE INTENTIONALLY BLANK

Practice Problems

Name _____

Find each product.

1) $2x(3x - 7)$

2) $-5x^3(6x^2 + 2y^2)$

3) $a^2b^4(a^2b - 3ab^2)$

4) $\frac{1}{3}mn^3\left(\frac{2}{5}m^3n - \frac{4}{3}m^2n^2\right)$

5) $5b(2b^3 - 4b^2 + 3b)$

6) $2x^3y(4x^3y - x^2y^2 + 3xy^3)$

7) $(x + 6)(x - 3)$

8) $(2y - 3)(y + 7)$

9) $(a - 2b)(4a + b)$

10) $(3m^2 + n)(m - 4n)$

Find each product.

11) $(x + 3)(x^2 - 3x + 2)$

12) $(3c - 2)(6c^2 + 5c - 11)$

13) $(m + n)(m + n)$

14) $(2p + 3q)(2p + 3q)$

15) $(3c - d)(3c - d)$

16) $(5g - 2h)(5g - 2h)$

17) $(x + y)(x - y)$

18) $(3a + 4b)(3a - 4b)$

19) $(2m^2 - 1)(2m^2 + 1)$

20) $\left(\dfrac{1}{2}x + \dfrac{1}{3}y\right)\left(\dfrac{1}{2}x - \dfrac{1}{3}y\right)$

Lesson 26: Factoring Polynomials

Greatest Common Factors

When factoring polynomials, the first thing to *always* check for is a greatest common factor (GCF) among all of the terms of the polynomial.

To find the greatest common factor among two or more numbers, write out the prime factorization of each number using the following procedure:

Divide each number by the smallest prime number it can be divided by. Starting with 2 if the number is even, or 3 if the number is odd, and then go up one prime number at a time (5, 7, 11, etc.) until you have found all of the prime numbers that can be multiplied together to create the number you started with.

---

example 1

Write out the prime factorization of 330 and 72.

$$
\begin{array}{cc}
330 & \qquad 72 \\
/\ \backslash & \qquad /\ \backslash \\
2 \times 165 & \qquad 2 \times 36 \\
\quad /\ \backslash & \qquad\quad /\ \backslash \\
\quad 3 \times 55 & \qquad\quad 2 \times 18 \\
\qquad /\ \backslash & \qquad\qquad /\ \backslash \\
\qquad 5 \times 11 & \qquad\qquad 2 \times 9 \\
& \qquad\qquad\quad /\ \backslash \\
& \qquad\qquad\quad 3 \times 3
\end{array}
$$

$$2 \cdot 3 \cdot 5 \cdot 11 = 330 \qquad\qquad 2 \cdot 2 \cdot 2 \cdot 3 \cdot 3 = 72$$

Once the prime factorization is written out, any common factors can be easily seen:

$$2 \cdot 3 \cdot 5 \cdot 11 \qquad\qquad 2 \cdot 2 \cdot 2 \cdot 3 \cdot 3$$

For the numbers 330 and 72, they each have prime factors of 2 and 3, so multiply the common prime factors together, and a greatest common factor of 6 is found.

---

example 2

Find the greatest common factor of $32x^2y$ and $24xy^2$.

The prime factorization of $32x^2y$ is:

$$2 \cdot 2 \cdot 2 \cdot 2 \cdot 2 \cdot x \cdot x \cdot y$$

The prime factorization of $24xy^2$ is:

$$2 \cdot 2 \cdot 2 \cdot 3 \cdot x \cdot y \cdot y$$

The common prime factors are 2, 2, 2, x, and y. Multiplying these together creates a greatest common factor of $8xy$.

To factor the binomial $32x^2y + 24xy^2$, find the GCF and then divide each term by that greatest common factor.

$$\frac{32x^2y}{8xy} = 4x \qquad \text{and} \qquad \frac{24xy^2}{8xy} = 3y$$

These are the terms that remain in the binomial, while the GCF is written out in front of this binomial:

$8xy(4x + 3y)$

To check your answer, multiply the binomial by the monomial using the distributive property:

$8xy(4x) + 8xy(3y)$

$32x^2y + 24xy^2$

And the answer is the original binomial.

Note: You can always check your answer when factoring by multiplying terms back together.

Factoring by Grouping

Remember, the first step in factoring is to always check for a GCF among all of the terms.

Next, if there are an even number of terms in the polynomial, you can *factor by grouping* by looking for a GCF among two terms at a time.

---

example 3

Factor: $3xy + 12x + 5x^2y + 10y$

First look for a greatest common factor among *all* of the terms.

In this example, there is no GCF for the entire polynomial.

Next, since there are an even number of terms, look for a GCF among each pair of terms:

$$(3xy + 12x) + (5x^2y + 10y)$$

The GCF of $3xy$ and $12x$ is $3x$, and the GCF of $5x^2y$ and $10y$ is $5y$.

Factor these common factors out of each binomial:

$$3x(y + 4) + 5y(x^2 + 2)$$

This is as far as the original polynomial can be factored.

This method of factoring by grouping can be used when factoring trinomials as well.

A **trinomial** is a polynomial with three terms. But an even number of terms is needed to factor by grouping. So turn the three terms into four terms by breaking up the middle term into two separate terms using the following method:

example 4

Factor: $2x^2 + 7xy + 3y^2$

Step 1) Multiply together the coefficients of the first and last terms of the polynomial.

$$\mathbf{2}x^2 + 7xy + \mathbf{3}y^2$$

$$2 \times 3 = 6$$

Step 2) Find two factors of this new number that happen to add up to the value of the middle coefficient.

$$2x^2 + \mathbf{7}xy + 3y^2$$

$$1 \times 6 = 6 \qquad \text{and} \qquad 1 + 6 = 7$$

Step 3) Rewrite the polynomial with two middle terms whose coefficients are the two numbers you found in step 2, but which have the same variable (or variables) as the original middle term.

$$2x^2 + 1xy + 6xy + 3y^2$$

Now you can factor by grouping:

$$(2x^2 + 1xy) + (6xy + 3y^2)$$

The GCF of $2x^2$ and xy is x, and the GCF of $6xy$ and $3y^2$ is $3y$.

$$x(2x + y) + 3y(2x + y)$$

Because the binomial that remains in parentheses is the *same* binomial, the answer can be written as the product of two binomials, the first one made up of each GCF that had been factored out before, and the second binomial being what remained after factoring:

$$(x + 3y)(2x + y)$$

Now the trinomial has been factored into two binomials, which if multiplied together (using the FOIL method) will result in the original trinomial.

Factoring Difference of Squares

Remember that $(x + y)(x - y) = x^2 - y^2$.

If you have a binomial with a perfect square minus a perfect square, the FOIL shortcut shown in Lesson 25 can be used in reverse.

example 5

Factor: $4a^2 - 9b^2$

$4a^2$ is a perfect square that comes from $2a \cdot 2a$

$9b^2$ is a perfect square that comes from $3b \cdot 3b$

So, using the FOIL shortcut for a difference of squares, $4a^2 - 9b^2$ can be factored into $(2a + 3b)(2a - 3b)$.

example 6

Factor: $5x^2 - 20y^2$

Neither $5x^2$ nor $20y^2$ are perfect squares, but when factoring, remember to *always check for a GCF first*.

A 5 can be factored out of this binomial:

$$5(x^2 - 4y^2)$$

The remaining binomial is now a difference of squares and can be factored into:

$$5(x + 2y)(x - 2y)$$

Practice Problems

Name _____

Factor completely. Be sure to factor out a greatest common factor when possible.

1) $3a^2 - 12a$

2) $-25y^3 + 10y$

3) $2a^2b + ab$

4) $2x^3 + 12x^2 - 8x$

5) $14c^3d^2 - 21c^2d + 35c$

6) $5g^4 - 7g^3 + g^2$

7) $4x^3 + 10xy^2 - 2x^2 + 6x$

8) $m^2 + 2m + 5m + 10$

9) $a^2 - 3a + 2a - 6$

10) $b^2 + 8b + 12$

11) $m^2 + 3m - 18$

12) $t^2 - 7t + 12$

Factor completely. Be sure to factor out a greatest common factor when possible.

13) $g^2 + 7g - 44$ 14) $d^2 + 17d + 52$ 15) $15x^2 - 16x + 4$

16) $6y^2 - 5y - 4$ 17) $6p^2 + 7p + 2$ 18) $4x^2 - 12xy + 9y^2$

19) $m^2 + 8mn + 16n^2$ 20) $25c^2 - 20cd + 4d^2$ 21) $2z^2 - 2z - 40$

22) $4x^2 + 12x + 8$ 23) $6x^2 + 15x - 9$ 24) $x^2 - y^2$

25) $c^2 - 4d^2$ 26) $4m^2 - 36n^2$ 27) $27m^2 - 12n^2$

Lesson 27: Solving Quadratic Equations

Polynomials are made up of monomials, each of which has a quality called the **degree** of that monomial. The degree of a monomial is the sum of the exponents of the variables in that monomial (or term). The degree of a *polynomial* is the highest degree of any term in the polynomial.

---

example 1

Find the degree of: $2x^4y^3 + 5x^2y + 6x$

Variables without an exponent written out have a power of 1:

$$2x^4y^3 + 5x^2y^1 + 6x^1$$

Add together the exponents in each term:

$$2x^4y^3 + 5x^2y^1 + 6x^1$$
$$\uparrow \qquad \uparrow \qquad \uparrow$$
$$(4 + 3 = \underline{7}) \,(2 + 1 = \underline{3})\,(\underline{1})$$

The largest degree of any of the terms is 7, so this is a polynomial of *degree 7*.

A **quadratic equation** is a second-degree polynomial of one variable that is set equal to zero, such as: $x^2 + 4x + 3 = 0$

One method for *solving* a quadratic equation is to use the **zero product property**:

> if $x \cdot y = 0$, either $x = 0$, $y = 0$, or they both equal zero.

In other words, the *only* way to multiply two (or more) terms together and get an answer of zero is if at least one of those terms is equal to zero.

So *when a polynomial is set equal to zero*, if that polynomial can be factored, then at least one of the factors must equal zero.

---

example 2

Solve for x: $x^2 + 4x + 3 = 0$

after factoring, the polynomial can be rewritten:

$$(x + 3)(x + 1) = 0$$

So either $x + 3 = 0$, or $x + 1 = 0$.

Now solve each of these new equations separately:

$$x + 3 = 0 \qquad\qquad x + 1 = 0$$
$$\underline{-3 \ -3} \qquad\qquad \underline{-1 \ -1}$$
$$x = -3 \qquad\qquad x = -1$$

This means there are two possible answers for the given quadratic equation.

---

example 3

Solve for x:

$$x^3 - 7x^2 = 18x$$

If the equation isn't already set equal to zero, move all of the terms to one side first:

$$x^3 - 7x^2 - 18x = 0$$

This is not a quadratic equation because it is a polynomial of degree 3, not degree 2, but there is a greatest common factor of x:

$$x(x^2 - 7x - 18) = 0$$

Once x is factored out, what remains is a polynomial of degree 2 that can be factored into:

$$x(x - 9)(x + 2) = 0$$

Now the zero product property can be used.

$$x = 0 \qquad\qquad x - 9 = 0 \qquad\qquad x + 2 = 0$$
$$\qquad\qquad \underline{+9 \ +9} \qquad\qquad \underline{-2 \ -2}$$

The three solutions are:

$$x = 0, \ x = 9, \ x = -2$$

Note: If a monomial (with a variable) can be factored out, like in the above example, *do not divide both sides by that monomial*:

$$\frac{x(x^2 - 7x - 18)}{x} = \frac{0}{x}$$

$$x^2 - 7x - 18 = 0$$

$$(x - 9)(x + 2) = 0$$

$$x - 9 = 0 \qquad\qquad x + 2 = 0$$

$$x = 9 \qquad\qquad x = -2$$

Dividing both sides by x in the beginning leads to two solutions. But this problem has *three* solutions. The solution $x = 0$ gets lost if you divide by x, so be sure to *only* factor when solving. This is very tempting to do, as shown in this next example.

---

example 4

Solve for x:

$$x^2 - 4x = 0$$

This quadratic equation can be factored into:

$$x(x - 4) = 0$$

Next, if both sides are divided by x:

$$\frac{x(x-4)}{x} = \frac{0}{x}$$

$$x - 4 = 0$$
$$+4 \;+4$$

$$x = 4$$

Remember, this is an example of the WRONG way to solve the quadratic equation. Solve by factoring and then using the zero product property, *not* by factoring and then dividing away any monomial terms with a variable.

When a quadratic equation cannot be solved by factoring, there is a second method that can be used.

---

example 5

Solve for x:

$$2x^2 + 7x + 4 = 0$$

First try to factor the polynomial by multiplying the coefficients of the first and last terms and finding factors that add up to the value of the middle coefficient.

$2 \times 4 = 8$ $\quad \Rightarrow \quad$ There are no two factors of 8 that add up to 7.

$$1 \times 8 = 8, \text{ but } 1 + 8 = 9$$
$$2 \times 4 = 8, \text{ but } 2 + 4 = 6$$

This polynomial is **prime** because it cannot be factored, but as a quadratic equation, it still has solutions that can be found by using the **quadratic formula**.

When you have a quadratic equation of the form $ax^2 + bx + c = 0$ (where $a \neq 0$), it can be solved by using the following formula:

$$x = \frac{-b \pm \sqrt{b^2 - 4ac}}{2a} \qquad \textbf{the quadratic formula}$$

Using this formula, the quadratic equation from the previous example can be solved:

1) First identify the value of a, b, and c:

$2x^2 + 7x + 4 = 0$, where $a = 2$, $b = 7$, and $c = 4$

Note: The polynomial must be set equal to zero before determining the value of a, b, and c.

2) Substitute the values of a, b, and c into the quadratic formula and simplify the expression:

$$x = \frac{-b \pm \sqrt{b^2 - 4ac}}{2a}$$

$$x = \frac{-(7) \pm \sqrt{(7)^2 - 4(2)(4)}}{2(2)}$$

$$x = \frac{-7 \pm \sqrt{49 - 32}}{4}$$

$$x = \frac{-7 \pm \sqrt{17}}{4}$$

At this point, if the number within the radical is not a perfect square, leave your answer in this form, which represents *two* solutions:

$$x = \frac{-7 + \sqrt{17}}{4} \qquad \text{and} \qquad x = \frac{-7 - \sqrt{17}}{4}$$

If the number within the radical is *negative*, then there is no real solution since you cannot take the square root of a negative number.

If the number within the radical *is* a perfect square, evaluate that square root and continue simplifying the expression.

Practice Problems

Name _____

Solve each equation. Remember to check your answer.

1) $x(x+4)=0$

2) $2c(c-5)=0$

3) $(b+2)(b-3)=0$

4) $(2a+3)(3a-1)=0$

5) $(3c+4)^2=0$

6) $\left(\dfrac{2f}{5}+4\right)^2=0$

7) $4h^2-20h=0$

8) $\dfrac{2}{3}b^2-\dfrac{4}{3}b=0$

9) $x^2+5x+4=0$

10) $y^2-y-12=0$

Solve each equation. Remember to check your answer.

11) $3a^2 - 11a - 4 = 0$

12) $8b^2 + 2b = 15$

13) $e^3 - 5e^2 - 14e = 0$

14) $4m^2 - 9 = 0$

Solve each equation using the quadratic formula.

15) $2x^2 + 9x + 4 = 0$

16) $m^2 + 3m = 2$

17) $3b^2 + 5b - 2 = 0$

18) $n^2 - 7n = -3$

19) $3a^2 + 2a = 3$

20) $y^2 = 5y - 2$

Lesson 28: Functions

<u>Functions</u>

A **function** is a special type of relation. A **relation** is a set of ordered pairs.

example 1

$$\{(2, 4), (1, -7), (5, 0), (3, -4)\}$$

The **domain** of a relation is the set of the first coordinates of the ordered pairs, and the **range** of a relation is the set of second coordinates of the ordered pairs.

From the above example:

domain = $\{2, 1, 5, 3\}$

range = $\{4, -7, 0, -4\}$

A relation can also be expressed as an equation, since ordered pairs can be created *from* an equation.

example 2

$y = 2x + 1$

An infinite number of ordered pairs can be created from this equation.

x	y
0	1
1	3
2	5
3	7
⋮	⋮

To be a **function**, the domain of a relation *cannot* repeat. In other words, when a relation is expressed as a list of ordered pairs, none of the *x* values can repeat.

example 3

$$\{(2, 7), (3, -4), (-8, 6), (4, 2)\}$$

This relation *is* a function because none of the values of the domain repeat.

173

| example 4 |

$\{(4, 0), (3, -2), (4, 7), (6, -3)\}$

This relation *is not* a function because of the 4 that repeats in the domain.

If a relation is graphed, it must be able to pass the **vertical line test** in order to be a function. If any points on the graph have the same *x*-value, a vertical line can be drawn that passes through these points. This means the relation is not a function. If there are no vertical lines that can intersect the graph at more than one point, then the relation passes the vertical line test and *is* a function.

relation 1

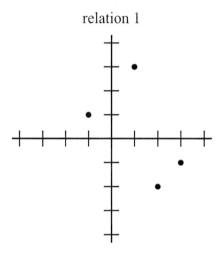

No vertical line can be drawn that passes through more than one point of the relation at a time.

⇒ It passes the vertical line test and is therefore a function.

relation 2

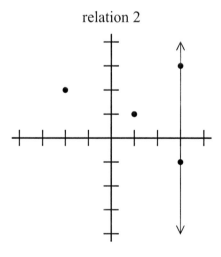

A vertical line can be drawn through more than one point of the relation at a time.

⇒ It does not pass the vertical line test and is therefore *not* a function.

example 5

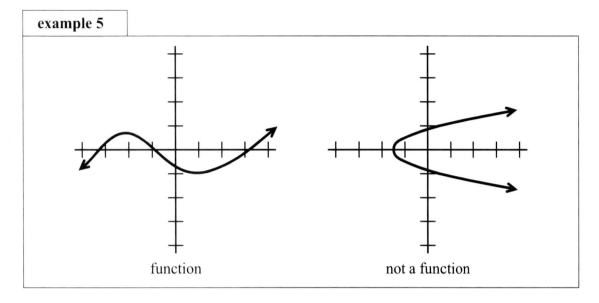

function   not a function

Function Notation

Equations that are functions can be written in **function notation**. The equation first must be solved in terms of y, such as the following:

$y = 2x + 1$

This relation is a function because no elements of the domain repeat, and if graphed, it would pass the vertical line test.

To rewrite this equation in function notation, replace y with $f(x)$, which stands for the "function of x", and is said "f of x".

$y = 2x + 1$ becomes $f(x) = 2x + 1$

Instead of substituting a value for x and solving for y, a number is substituted for x to get a value for the *function*.

example 6

Find the value of $f(2)$ for the given function.

$f(x) = 2x + 1$

To solve, replace x with a 2:

$f(2) = 2(2) + 1$

$f(2) = 4 + 1$

$f(2) = 5$

A function doesn't have to be linear.

---

example 7

Find the value of $f(4)$ for the given function.

$$f(x) = 3x^2 + 2x + 6$$

To solve, replace x with a 4:

$$f(4) = 3(4)^2 + 2(4) + 6$$

$$f(4) = 3(16) + 8 + 6$$

$$f(4) = 48 + 14$$

$$f(4) = 62$$

Practice Problems

Name _____

State whether each relation is a *function* or *not a function*.

1) $\{(2, -1), (3, 4), (-1, 5), (2, 5)\}$

2) $\{(4, -2), (6, -2), (-1, -2,), (0, -2)\}$

3) $\{(1, 5), (1, -3), (1, 4), (1, -6)\}$

4) $\{(-5, 2), (0, 5), (3, -2), (4, 1)\}$

5) $x^2 + y^2 = 4$

6) $y = 2x + 3$

7)
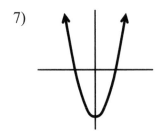

8)
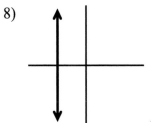

Given $f(x) = 3x + 1$, $g(x) = x^2 - 2x + 3$, and $h(x) = x^3 - 7$, find each value.

9) $f(3)$

10) $g(5)$

11) $h(-2)$

12) $f(-5)$

13) $g(0)$

14) $h(4)$

15) $f(13)$

16) $g\left(\dfrac{1}{2}\right)$

17) $h(-1)$

18) $f(y)$

19) $g(x + 2)$

20) $h(a)$

Lesson 29: Systems of Equations

Two or more linear equations involving the same variables are called a **system of equations**.

Solving a system of linear equations means finding the value of each variable that is true in each equation.

There are three methods of solving a system of equations.

Method 1 - Graphing

Graph each linear equation on the same coordinate plane.

If there is *one solution* to the system of equations, the lines will intersect, and the x- and y-value of that point of intersection is the one solution what will work in each of the equations.

---

example 1

Solve by graphing: $x + y = 4$
                       $x - y = 4$

$\Rightarrow$ one solution: $(4, 0)$

If there is *no solution* to the system of equations, the lines will be parallel, meaning there are no points that the lines have in common, or in other words, no solutions that work in each of the equations.

---

example 2

Solve by graphing: $2x - 3y = 6$
                       $2x - 3y = -6$

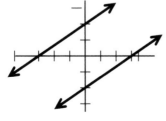

$\Rightarrow$ no solution

If there are an *infinite number of solutions*, the two lines will lay one on top of the other, meaning every solution for the first equation works in the second equation.

---

example 3

Solve by graphing: $x + 2y = 4$
                         $2x + 4y = 8$

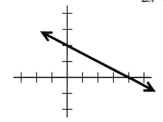

$\Rightarrow$ infinite number of solutions

Method 2 – Substitution

Solve one of the equations for one variable, and then replace that variable in the second equation with the expression it was found to equal in the first equation.

---

example 4

Solve the system of equations: $3x + y = 7$
                                     $4x - 2y = 6$

Hint: Look to see if either equation has a variable with a coefficient of 1 and solve that equation in terms of that variable.

$3x + y = 7$
$-3x \qquad -3x$

$y = 7 - 3x$

Now replace y in the second equation with $7 - 3x$:

$4x - 2y = 6$

$4x - 2(7 - 3x) = 6$

$4x - 14 + 6x = 6$

$10x - 14 = 6$
$\quad\; +14 \; +14$

$10x = 20$

$\dfrac{10x}{10} = \dfrac{20}{10}$

$x = 2$

Now substitute 2 in for x in the rearranged first equation:

$$y = 7 - 3x$$

$$y = 7 - 3(2)$$

$$y = 7 - 6$$

$$y = 1$$

The solution to this system of equations is (2, 1).

Method 3 – Elimination

In a system of equations where the coefficients of x or y in both equations are additive inverses, add the equations together.

example 5

Solve the system of equations: $2x - 3y = 7$
$4x + 3y = 5$

The coefficients of both y terms are additive inverses (–3 and 3), so add the equations together to get a new equation that is easily solved for x:

$$2x - 3y = 7$$
$$(+) \quad 4x + 3y = 5$$
_____

$$6x + 0y = 12$$

$$6x = 12$$

$$\frac{6x}{6} = \frac{12}{6}$$

$$x = 2$$

Now substitute 2 in for x into either of the original equations and solve for y:

$$2x - 3y = 7$$

$$2(2) - 3y = 7$$

$$4 - 3y = 7$$
$$-4 \qquad -4$$

$$-3y = 3$$

$$\frac{-3y}{-3} = \frac{-3}{-3}$$

$$y = -1$$

The solution to the system of equations is (2, –1).

If the coefficients of x and y in a system of equations are *not* additive inverses, multiply one or both equations by a number that will give additive inverse coefficients for x or y.

---

example 6

Solve the system of equations: $2x - 5y = -10$
                                                         $4x + 3y = 6$

If the top equation is multiplied by –2:

$$(-2)(2x - 5y) = (-10)(-2)$$
$$4x + 3y = 6$$

A new equation results with a coefficient of x that is an additive inverse to the coefficient of x in the second equation, and now the equations can be added together:

$$-4x + 10y = 20$$
$$(+)\quad 4x + 3y = 6$$
$$\overline{}$$
$$0x + 13y = 26$$

$$13y = 26$$

$$\frac{13y}{13} = \frac{26}{13}$$

$$y = 2$$

Substitute 2 in for y in any of the original equations and solve for x, and you get a solution to the system of equations of $(0, 2)$.

---

example 7

Solve the system of equations: $5x + 3y = -6$
                                                         $3x + 5y = -10$

Sometimes you will need to multiply both equations by a number in order to make the elimination method work.

In this example, either multiply the first equation by –3 (or 3) and the second equation by 5 (or –5) to eliminate the x terms, or multiply the first equation by 5 (or –5) and the second equation by 3 (or –3) to eliminate the y terms:

$$(-3)(5x + 3y) = (-6)(-3) \quad \rightarrow \quad -15x - 9y = 18$$
$$(5)(3x + 5y) = (-10)(5) \quad \rightarrow \quad 15x + 25y = -50$$

Now elimination can be used to create an equation that can be easily solved for y. To finish the problem, follow all of the steps mentioned above to find a solution of $(0, -2)$.

Practice Problems

Name _____

Solve each system of equations by graphing.

1) $x + y = 3$
    $x - y = -1$

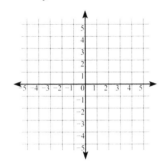

2) $2x - y = -3$
    $x + y = -3$

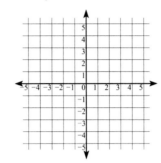

3) $3x - y = -4$
    $3x - y = 0$

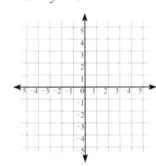

4) $2x - y = 3$
    $4x - 2y = 6$

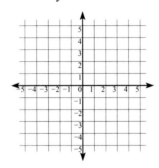

Solve each system of equations using substitution.

5) $x - 4y = 8$
    $2x - 8y = 16$

6) $3x - y = 4$
    $2x - 3y = -9$

7) $x + y = 2$
    $x - y = -4$

8) $2x - 6y = -24$
    $x - 3y = 18$

Solve each system of equations using elimination.

9) $x + 2y = 8$
    $x - 2y = 4$

10) $2x + y = -4$
    $2x - 5y = -4$

11) $2x + 3y = 6$
    $2x + 3y = -3$

12) $x - 2y = -3$
    $4x + 3y = -12$

Solve each system of equations using any method.

13) $2x - 5y = 13$
    $3x + y = 11$

14) $x + y = 2$
    $x - y = 0$

15) $2x - y = -3$
    $x - 3y = 11$

16) $x + 3y = 4$
    $2x + 6y = 8$

Lesson 30: Problem Solving

One of the most difficult tasks in math is to set up an equation to be solved before even attempting to solve it.

Setting Up Algebraic Expressions

An **algebraic expression** is one or more algebraic terms in a phrase. These terms can include variables, constants, and operating symbols (plus signs, minus signs, etc.). An expression does *not* have an equals sign, so it is not an equation, although it can be part of an equation.

The key to setting up an expression is knowing how variables relate to each other — what is being multiplied together or divided into, or what is being added together or subtracted. Knowing what operations are taking place is just as important as knowing what each variable represents.

---

example 1

Kristen was x inches tall a year ago. Since then, she has grown taller and is now y inches tall. Which of the following represents the number of inches Kristen grew during the past year?

A. $x + y$
B. $x - y$
C. $y - x$
D. $y \div x$

Once again, the first step in setting up this algebraic expression is choosing the correct operation. This problem is asking about the difference in Kristen's height from last year to this year. A *difference* is an answer that comes from *subtraction*, so we need to subtract last year's height from this year's height.

The answer is C.

Balancing Equations

An algebraic expression becomes part of an equation when it is set equal to another expression. The symbol for showing that two expressions have the same value is the *equals sign* ($=$).

While this may seem obvious, a lot of people forget that both sides must remain equal at all times. When you are trying to solve an equation, variables and constants will be added, subtracted, multiplied, and divided as you try to get a variable all alone on one side and see what it ends up equaling. While doing so, you need to remember that you can do anything to an equation, as long as you do it to *all* terms on *both* sides of the equals sign. (Note: The one exception is you can't divide by zero.)

In this way, an equation is like a scale. As long as it is in balance, both sides are equal. If you make any changes to one side, you must make those same changes to the other side.

---

example 2

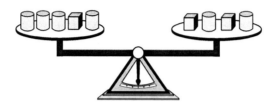

Which of the following shows the relationship between the weights of one cylinder and one cube?

A. One cube weighs the same as two cylinders.
B. One cube weighs the same as four cylinders.
C. One cylinder weighs the same as two cubes.
D. One cylinder weighs the same as four cubes.

The objects on each side of the scale can be thought of as variables. We don't know how much each one weighs, but there is a relationship between them all because they are balanced, just like an equation would be. If we let x represent the cylinders, and y represent the cubes, we can write an equation like this:

$$4x + y = 2x + 2y$$

Now let's get all the x's on one side and all the y's on the other side to see how they relate to each other:

$$4x + y = 2x + 2y$$
$$ -y -y$$
$$4x = 2x + y$$
$$-2x -2x$$
$$2x = y$$

This means that two of the cylinders is equal to one of the cubes.

The answer is A.

Problem Solving

When given a word problem where you have to set up the equation to solve, there are four steps to follow:

1) Explore the problem and define a variable.

   This means you should read through the problem carefully and see what information is provided. Also, *make sure you know what you are being asked to find*. This is usually the question at the end of the **problem**.

2) Translate English into Algebra to set up the equation to be solved.

3) Solve the equation.

4) Examine the solution to make sure it makes sense and *check your work*.

Some students are so relieved to have set up *any* equation that they don't check to make sure that they actually answered the questioned being asked.

It's also important to check your math and make sure you didn't make any careless arithmetic errors while solving your equation. Make sure minus signs didn't disappear or appear incorrectly and that simple addition and subtract mistakes weren't made.

Another good way to double-check your answer is to make sure the **units of measurement** of your answer match the units of the original question. For example, if you're asked to find a distance your answer shouldn't be in units of time.

---

example 3

The president of the science club brought 134 cans of juice on a field trip. Each person on the trip received 3 cans of juice, and there were 20 extra cans. Which equation could be used to find n, the number of people who went on the field trip?

A. $134 = \dfrac{n}{3} - 20$

B. $134 = \dfrac{n}{3} + 20$

C. $134 = 3n - 20$

D. $134 = 3n + 20$

In all the steps outlined above for how to solve a problem, Step 1 is taken care of for us because we're told that the variable n will represent the number of people who went on the field trip.

Step 2 is to set up the equation to be solved, and that is what this question is looking for. If each person is to receive 3 cans, the number of cans per person can be represented by multiplying *n* by 3, not dividing *n* by 3 (like in Answers A and B).

Now that we know that 3*n* should be part of the equation, should the remaining 20 cans be added to that or subtracted? Since they are extra cans, they will be added so that *all* cans are represented by either the ones given to the students or the ones that were extra.

The answer is D.

---

example 4

Marisa saved $500 to spend on a vacation. She will spend about $45 per day on her vacation, and she must have $70 left to pay for her bus ride home. Which of the following inequalities **best** represents the possible number of days, *d*, Marisa could be on vacation?

A. $500 – ($45 *d*) ≥ $70
B. $500 – ($45 *d*) ≤ $70
C. $500 – ($70 *d*) ≥ $45
D. $500 – ($70 *d*) ≤ $45

The key phrase in this problem is "$45 per day" which tells us that ($45 *d*) should be part of the inequality. The word "per" can sometimes refer to division (when it is part of a rate), but in this case it refers to multiplication.

The other thing to consider when setting up this inequality is that Marisa needs at least $70 to remain by the end of her trip. With inequalities, the phrase "at least" is another way of saying "greater than or equal to" (while "at most" means "less than or equal to"). This means the inequality symbol used in this problem will be the "≥" symbol.

The answer is A.

---

example 5

In John's homeroom, $\frac{1}{3}$ of the students walk to school and $\frac{1}{4}$ come by car. The remaining 15 come by school bus. How many students are in his homeroom?

A. 48
B. 24
C. 36
D. 21

This problem isn't asking for the equation or inequality needed to solve. It is looking for the actual solution, so we'll go through all four steps for problem solving here.

Step 1) First, explore the problem and find the numbers it gives that will be used when you set up an equation: $\frac{1}{3}$, $\frac{1}{4}$, and 15. Next, determine what the variable will be. *What is unknown in this problem?* This is usually the question at the end of the problem: "How many students are in his homeroom?" So the unknown number is the total number of students, which you can call x, or n for "number", or any letter you want.

Step 2) Now create the equation you will solve. One-third of all the students who walk, plus one-fourth of all the students who travel by car, plus the remaining 15 students add up to the total number of students in the class. The key word here is "of", which refers to multiplication. One-third *of* all the students means one-third *times* the total number of students. This translates into the following equation:

$$\frac{1}{3}x + \frac{1}{4}x + 15 = x$$

Step 3) Once you have an equation, solve it:

$$\frac{1}{3}x + \frac{1}{4}x + 15 = x$$

$$\frac{4}{12}x + \frac{3}{12}x + 15 = x$$

$$\frac{7}{12}x + 15 = \frac{12}{12}x$$

$$-\frac{7}{12}x \qquad\qquad -\frac{7}{12}x$$

$$15 = \frac{5}{12}x$$

$$\frac{12}{5} \times 15 = \frac{5}{12}x \times \frac{12}{5}$$

$$\frac{180}{5} = x$$

$$x = 36$$

There are 36 student in the class.

Step 4) When you solve the equation you aren't done yet. Check your work, make sure your answer makes sense, *and* make sure you've answered the problem. One way to check your work is to substitute your answer into the equation you came up with and make sure it works:

$$\frac{1}{3}x + \frac{1}{4}x + 15 = x$$

$$\frac{1}{3}(36) + \frac{1}{4}(36) + 15 = (36)$$

$$\frac{36}{3} + \frac{36}{4} + 15 = 36$$

$$12 + 9 + 15 = 36$$

$$36 = 36$$

This only proves that you solved *your* equation correctly. What if your equation wasn't the correct one to set up in the first place? This is why you need to make sure the answer is not only mathematically correct, but that it makes sense.

Also, was the question asked in the problem actually answered? Usually you can define your variable to represent *exactly* what the problem is asking for, but not always. Or in some problems, you'll be asked to find two or more values that are easy to find once you've solved your equation, but you have to remember to go back and do that and not stop once you've solved for *x* (or whatever variable you use).

Practice Problems

Name _____

1) If a number is decreased by 8, and the difference is multiplied by 3, the result is the same as twice the original number then decreased by 4. Find the number.

2) To raise money, the school band has decided to sell T-shirts. The shirts cost $5.00 each, plus $1.00 per shirt to have the school mascot printed on the front. If the shirts are sold for $11.00 each, how many shirts will need to be sold to make a profit of at least $300?

3) A cell phone company charges a monthly fee of $8.95 for 500 text messages, and $0.10 for each additional text message. If a customer sent or received 545 text messages, how much should he be charged?

4) In Ron's homeroom, one-third of the students walk to school, half of the students take the bus, and the rest get dropped of by their parents. If 4 students are dropped off at school their parents, how many students are in Ron's homeroom?

5) When the lesser of two consecutive even integers is multiplied by 5, the result is the same as 4 times the greater number. Find the integers.

6) The length of a rectangle is 2 less than 3 times the width. If the perimeter is 20, find the dimensions of the rectangle.

7) 30% of the students at a school play on a sports team. If there are 288 student-athletes, how many students are in the entire school?

8) Samantha is 6 years older than her brother. Four years ago she was 3 times his age, and last year she was twice his age. How old are Samantha and her brother?

9) There are 10% more students in the freshman class than in the sophomore class. Combined there are 630 students in both classes. How many sophomores are there?

10) Karen bought a jacket that costs $82.95 with tax. If the sales tax rate is five percent, what was the price of the jacket before tax?

Lesson 31: Weighted Averages

To calculate an average, add up all of the terms to be averaged and then divide that value by the total number of terms.

With a **weighted average**, a different problem-solving approach is needed because some of the values to be averaged must be weighted more than other values.

For example, if you are in a class where every test is worth 20% of your grade, but every quiz is worth 10% of your grade, you must *weigh* your test scores twice as much as your quiz scores when calculating your grade for the course.

There are typically three types of weighted average problems you may run into:

1) distance, rate and time problems

2) problems involving percents

3) problems with items of varying value (or varying *weight*)

Each of these types of weighted average problems can all be solved the same exact way.

example 1

A train leaves the station heading east at 60 miles per hour. At the same time, a second train leaves the station heading west at 80 miles per hour. How long will it take for these two trains to be 350 miles apart?

The method for solving a weighted average problem generally has four steps:

Step 1) Create a table with three columns for numbers, variables, or algebraic expressions, and with two or three rows (depending on the number of items involved in the problem, two items in this case — two trains).

In a distance, rate, and time problem, the formula $d = r \cdot t$ will be used, but for weighted average problems, use $r \cdot t = d$ instead, and label the table in the same way, with the first column for rate, the second column for time, and the third column for distance.

	r	t	d
train 1			
train 2			

Step 2) Fill in all of the information provided by the problem into the first two columns of the table. Any unknowns (variables) should appear in one of the first two columns as well.

	r	t	d
train 1	60	x	
train 2	80	x	

The time spent traveling is the same for both trains since they each left the station at the same time.

Step 3) Multiply across each row (column 1 × column 2) to create a value or algebraic expression to be entered into column 3.

$$60 \cdot x = 60x \quad \text{and} \quad 80 \cdot x = 80x$$

	r	t	d
train 1	60	x	60x
train 2	80	x	80x

Step 4) Use the expressions from the third column, along with any information from the original problem that was not put in the table, to set up an equation to solve.

$$60x + 80x = 350$$

$$140x = 350$$

$$\frac{140x}{140} = \frac{350}{140}$$

$$x = 2.5$$

The trains will be 350 miles apart 2½ hours after they leave the station.

example 2

How much pure orange juice must be added to 2 quarts of a drink that contains 10% orange juice to make a mixture that is 50% orange juice?

Step 1) In a problem like this, the percentages are labels for each drink, but are *also* numerical information that will be used in the problem.

% juice	no. of quarts	quarts of OJ
10% juice		
100% juice (pure OJ)		
50% juice		

This table has three rows because there are three items: pure orange juice, a 10% orange juice drink, and a 50% orange juice drink.

Step 2)

% juice	no. of quarts	quarts of OJ
10% juice	2	
100% juice (pure OJ)	x	
50% juice	$x + 2$	

Under **number of quarts**, there is an unknown amount of pure orange juice, 2 quarts of the 10% drink, and if they are being added together, then the amount of the 50% drink will be $(x + 2)$.

Step 3) Multiply across.

% juice	no. of quarts	quarts of OJ
10% juice	2	$(.10)2$
100% juice (pure OJ)	x	$(1.00)x$
50% juice	$x + 2$	$(.50)(x + 2)$

Step 4) Set up the equation to solve from column 3.

$$x + 0.2 = 0.5(x + 2)$$

$$x + 0.2 = 0.5x + 1$$
$$-0.5x \qquad -0.5x$$

$$0.5x + 0.2 = 1$$
$$ -0.2 \; -0.2$$

$$0.5x = 0.8$$

$$\frac{0.5x}{0.5} = \frac{0.8}{0.5}$$

$$x = 1.6$$

1.6 quarts of pure orange juice must be added to 2 quarts of a 10% OJ drink to get a mixture that is a 50% OJ drink.

Note: When setting up the equation you will solve, the terms of the equation come from column 3. If there are two rows in your table, the two terms in column 3 will typically be added together and set equal to a number from the problem that wasn't put into the table (like in **example 1**). If there are three rows in your table, the terms (in column 3) from rows 1 and 2 will typically be added and set equal to the term in row 3 (like in **example 2**).

example 3

A group going to a football game bought 23 tickets. Adult tickets cost \$15 while student tickets cost \$8. The group spent a total of \$268. How many of each kind of ticket were purchased?

Step 1)

	no. of tickets	price per ticket	total cost
adult tickets			
student tickets			

This table has two rows because there are two items involved: adult tickets and student tickets.

Step 2)

	no. of tickets	price per ticket	total cost
adult tickets	x	\$15	
student tickets	$23 - x$	\$8	

The number of each kind of ticket is unknown, so it doesn't matter which one is called x, but whichever one is, since the total is known, the number of the other kind of ticket is that total minus x.

Step 3)

	no. of tickets	price per ticket	total cost
adult tickets	x	\$15	$15x$
student tickets	$23 - x$	\$8	$(23 - x)8$

Step 4) The amount spent on adult tickets and the amount spent on student tickets will add up to the total amount spent on all of the tickets.

$$15x + 8(23 - x) = 268$$

$$15x + 8(23) + 8(-x) = 268$$

$$15x + 184 - 8x = 268$$

$$7x + 184 = 268$$
$$-184\ -184$$

$$7x = 84$$

$$\frac{7x}{7} = \frac{84}{7}$$

$$x = 12$$

12 adult tickets were bought, along with 23 minus 12, or 11 student tickets.

Practice Problems

Name _____

1) Penny has 4 more dimes than the number of quarters that she has. If she is carrying a total of $2.50, how many of each coin does she have?

2) How much water must be added to 6 liters of a 40% acid solution to obtain a 30% acid solution?

3) Ben's car radiator has 6 quarts of a mixture that is 40% antifreeze. How many quarts of the mixture should be drained and replaced with pure antifreeze to obtain a mixture that is 50% antifreeze?

4) A ship sailing from Boston, Mass. to Miami, Florida takes 50 hours to complete its trip. A jet traveling 570 miles per hour faster than the ship can reach Miami from Boston in 2.5 hours. How fast is the ship traveling?

5) Tina's lemonade stand sells cups of lemonade for $0.50 and cookies for $1.25. If Tina sold 37 items one day and made $29.00, how many cups of lemonade did she sell?

6) Two trains leave the station at the same time, but in opposite directions, with one heading north and the other traveling south. If the first train travels at 55 miles per hour while the second train travels at 70 miles per hour, in how many hours will they be 500 miles apart?

7) A boat can travel 48 miles down-stream, with a current of 2 miles per hour, in 4 hours. The return trip, going against the current of 2 miles per hour, takes 6 hours. What is the speed of the boat in still water with no current?

8) A manufacturer has an alloy that is 40% steel and an alloy that is 80% steel. How many kilograms of the 80% alloy should be combined with 4 kilograms of the 40% alloy to create an alloy that is 50% steel?

National Council of Teachers of Mathematics
Standards for School Mathematics

Geometry and Measurement Standards for Grades 6–8
Expectations

Instructional programs from prekindergarten through grade 12 should enable all students to—	In grades 6–8 all students should—	The following lessons correspond to each expectation—
Analyze characteristics and properties of two- and three-dimensional geometric shapes and develop mathematical arguments about geometric relationships	• precisely describe, classify, and understand relationships among types of two- and three-dimensional objects using their defining properties; • understand relationships among the angles, side lengths, perimeters, areas, and volumes of similar objects; • create and critique inductive and deductive arguments concerning geometric ideas and relationships, such as congruence, similarity, and the Pythagorean relationship.	• Lessons 42 and 45 • Lesson 38 • Lessons 35 and 38
Specify locations and describe spatial relationships using coordinate geometry and other representational systems	• use coordinate geometry to represent and examine the properties of geometric shapes; • use coordinate geometry to examine special geometric shapes, such as regular polygons or those with pairs of parallel or perpendicular sides.	• Lesson 42 • Lesson 42
Apply transformations and use symmetry to analyze mathematical situations	• describe sizes, positions, and orientations of shapes under informal transformations such as flips, turns, slides, and scaling; • examine the congruence, similarity, and line or rotational symmetry of objects using transformations.	• Lesson 39 • no matching lessons
Use visualization, spatial reasoning, and geometric modeling to solve problems	• draw geometric objects with specified properties, such as side lengths or angle measures; • use two-dimensional representations of three-dimensional objects to visualize and solve problems such as those involving surface area and volume; • use visual tools such as networks to represent and solve problems; • use geometric models to represent and explain numerical and algebraic relationships; • recognize and apply geometric ideas and relationships in areas outside the mathematics classroom, such as art, science, and everyday life.	• no matching lessons • Lesson 45 • no matching lessons • no matching lessons • no matching lessons
Understand measurable attributes of objects and the units, systems, and processes of measurement	• understand both metric and customary systems of measurement; • understand relationships among units and convert from one unit to another within the same system; • understand, select, and use units of appropriate size and type to measure angles, perimeter, area, surface area, and volume.	• Lesson 40 • Lessons 40 and 41 • Lessons 32, 40, 43, 44, 46, and 47
Apply appropriate techniques, tools, and formulas to determine measurements	• use common benchmarks to select appropriate methods for estimating measurements; • select and apply techniques and tools to accurately find length, area, volume, and angle measures to appropriate levels of precision; • develop and use formulas to determine the circumference of circles and the area of triangles, parallelograms, trapezoids, and circles and develop strategies to find the area of more-complex shapes; • develop strategies to determine the surface area and volume of selected prisms, pyramids, and cylinders; • solve problems involving scale factors, using ratio and proportion; • solve simple problems involving rates and derived measurements for such attributes as velocity and density.	• Lesson 40 • Lessons 32, 40, 44, and 47 • Lessons 43 and 44 • Lessons 46 and 47 • Lesson 38 • no matching lessons

Lesson 32: Angle Measurement

Angle Measurement

To define an angle, something called a **ray** must first be defined. Start with a point (a **point** is a location in space that has no dimensions) and draw a line away from that point in one direction only. (This is not a geometric line because a **line** extends indefinitely in two directions.)

What you now have is a ray — a set of an infinite number of points that extends forever in one direction from one endpoint.

Now, without moving the endpoint (point *A*), rotate the ray in any direction (usually counterclockwise):

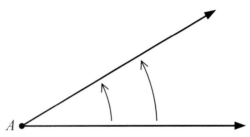

This creates two rays with the same endpoint (also called the vertex), which is known as an **angle**.

An angle can be labeled in several ways, and you should be familiar with each of them.

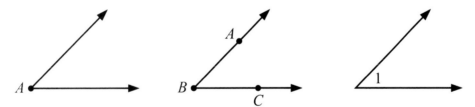

The symbol for an angle is $\angle$.

As shown above, an angle can be labeled by the vertex of the angle ($\angle A$), or an angle can be defined by three points on the angle, one point on each ray and the vertex ($\angle ABC$). When written this way, the vertex point is always the middle letter in the label of the angle. An angle can also be labeled with a number ($\angle 1$).

Be prepared to see angles labeled in any of these three ways.

The measure of an angle is the size of the "opening" of an angle. But this measurement is not a *distance* because the two rays that make up the angle get farther apart the more you move away from the vertex, as shown on the next page.

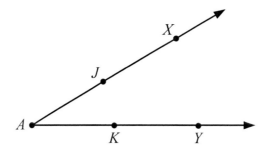

In ∠A, points X and Y are farther apart than the distance between points J and K, but the angle measurement is the same. Instead of *distance* for angle measurement, the "opening" of the angle is measured in **degrees**.

The best way to understand degree measurement is to understand a circle. There are 360 degrees (or 360º) in a circle, or in other words, if you rotate a ray through one complete revolution to make an angle, it would be an angle of 360º.

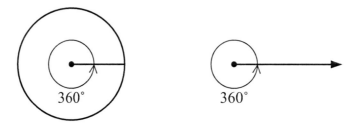

If you create an angle by rotating a ray one-fourth of the way around one full revolution, it would be an angle that is one-fourth of 360º, or a 90º angle (called a **right angle**).

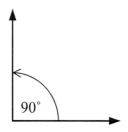

It is very important to have a basic understanding of the size of various angles:

In the above figures, the measure of angle A is 45º (this is usually written as m∠A = 45º), while m∠B = 90º, and m∠C = 180º.

example 1

What is the best estimate of the angle between the two hands on a clock at 5:10?

A. 45º

B. 30º

C. 90º

D. 60º

This problem brings up another useful way to remember the various sizes of angles: by using a clock.

A clock can be thought of as a circle divided into twelve slices, and since there are 360º in a circle, each slice would have an angle of 30º (360 ÷ 12 = 30).

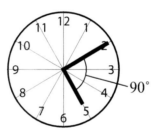

At 5:10, the angle between the hour hand and the minute hand is about 90º, which can be seen by counting how many slices are contained between the two hands of the clock (3 slices × 30º = 90º). Or, you could have noticed that the angle formed by the hands of the clock looks like a right angle.

The answer is C.

Some additional vocabulary is also useful to know about angles:

Angles with a measure less than 90º are called **acute**.

Angles with a measure greater than 90º but less than 180º are called **obtuse**.

An angle with a measure of exactly 90º is called a **right angle**.

An angle with a measure of exactly 180º is called a **straight angle**.

Complementary and Supplementary Angles

If the sum of the measures of two angles (and *only* two angles) is 90º, the angles are called **complementary angles**.

If the sum of the measures of two angles (and *only* two angles) is 180º, the angles are called **supplementary angles**.

These angles can be **adjacent angles** (meaning they have the same vertex and have a common side) or they don't have to be anywhere near each other.

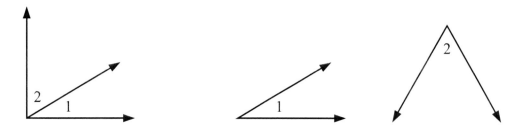

If m∠1 + m∠2 = 90º, then ∠1 and ∠2 are complementary angles.

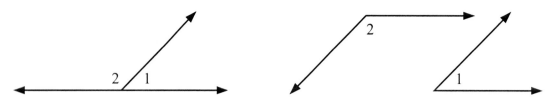

If m∠1 + m∠2 = 180º, then ∠1 and ∠2 are supplementary angles.

Note: Angles with a measure greater than or equal to 90º have no complement (a **complement** is an angle with a measure equal to the difference between 90º and the measure of another angle), and angles with a measure greater than or equal to 180º have no supplement (a **supplement** is an angle with a measure equal to the difference between 180º and the measure of another angle).

The adjacent supplementary angles shown above are also an example of a **linear pair** — adjacent angles whose non-common sides make a **line**.

Supplementary angles can also be created by the intersection of two lines.

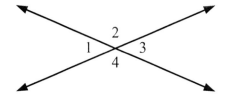

∠1 and ∠2 are supplementary angles, as are ∠2 and ∠3, ∠3 and ∠4, and ∠1 and ∠4.

(Each pair of angles opposite each other when two lines intersect are called **vertical angles**, such as ∠1 and ∠3, or ∠2 and ∠4.)

Practice Problems

Name _____

Estimate the degree measure of each of the following angles.

1)
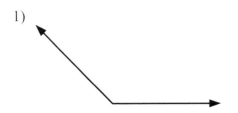

2)

3)
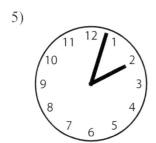

4)
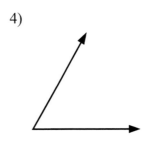

Estimate the measure of the angle formed by the hour and minute hand on each of the following clocks.

5)
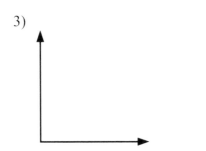

6)
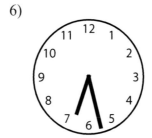

7)
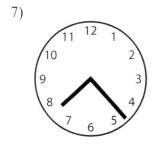

8)
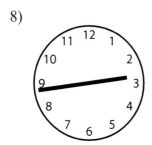

Refer to the diagram below for questions 9–12.

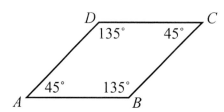

9) Which pair of angles are complementary?

10) Which pair of angles are supplementary?

11) Which pair of angles are acute?

12) Which pair of angles are obtuse?

Find the complement and supplement of each angle measure given.

13) 15°   14) 90°

15) 195°   16) 30°

17) 45°   18) 95°

19) 135°   20) 180°

21) An angle measures 26° more than its complement. Find the measures of the two angles.

22) An angle measures 60° more than its supplement. Find the measures of the two angles.

Lesson 33: Angles Formed by Transversals of Coplanar Lines

When two lines intersect, we've already seen how pairs of angles called vertical angles and linear pairs are created, and how linear pairs are supplementary angles.

A line that intersects *two or more lines* is called a **transversal**, and many pairs of angles are created.

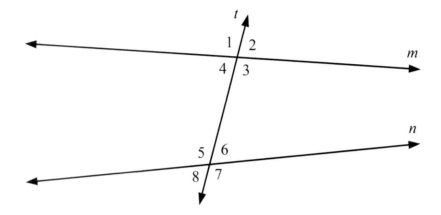

Above, lines *m* and *n* are intersected by line *t*, (the transversal), which creates 8 angles.

In this situation, any two angles that occupy a similar position in the two sets of four angles are called **corresponding angles**. In the diagram above, the corresponding angles are: $\angle 1$ and $\angle 5$, $\angle 2$ and $\angle 6$, $\angle 3$ and $\angle 7$, and $\angle 4$ and $\angle 8$.

A special situation arises if a transversal intersects two *parallel* lines.

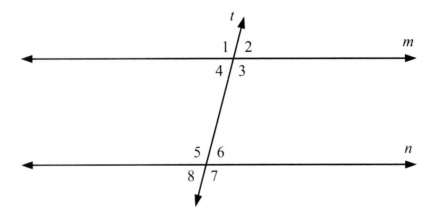

In this case, corresponding angles are *congruent*, which means they have the same measure.

The symbol for congruence is "≅", so $\angle 1 \cong \angle 5$, $\angle 2 \cong \angle 6$, $\angle 3 \cong \angle 7$, and $\angle 4 \cong \angle 8$.

Vertical angles are also congruent: $\angle 1 \cong \angle 3$, $\angle 2 \cong \angle 4$, $\angle 5 \cong \angle 7$, and $\angle 6 \cong \angle 8$. (This is true with *any* vertical angles, not just when parallel lines are involved.)

Interior angles are the angles on the inside of the parallel lines, and interior angles on opposite sides of the transversal (but do not form a linear pair) are called **alternate interior angles**, which also happen to be congruent: $\angle 4 \cong \angle 6$ and $\angle 3 \cong \angle 5$.

Exterior angles are the angles on the outside of the parallel lines, and exterior angles on opposite sides of the transversal (but do not form a linear pair) are called **alternate exterior angles**, which are also congruent: $\angle 1 \cong \angle 7$ and $\angle 2 \cong \angle 8$.

example 1

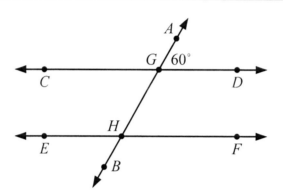

In the figure above, lines *CD* and *EF* are parallel. What is the measure, in degrees, of $\angle BHF$?

When a line crosses two parallel lines, eight angles are formed. If just one of the measures for any of those eight angles is given, finding the other seven is easy:

$\angle AGD$ and $\angle CGH$ are vertical angles, so they have the same measure of 60°. $\angle AGD$ and $\angle EHB$ are alternate exterior angles, so they have the same measure as well. And, $\angle CGH$ and $\angle GHF$ are alternate interior angles, so they also have the same measure:

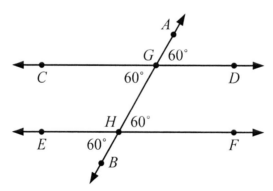

∠*AGD* and ∠*AGC* are a linear pair, so those angles are supplementary. If ∠*AGD* has a measure of 60º, then ∠*AGC* must have a measure of 120º. Using the same methods as above, ∠*HGD* must also have a measure of 120º because it and ∠*AGC* are vertical angles. And since ∠*HGD* and ∠*EHG* are alternate interior angles, they both have the same measure, as do ∠*AGC* and ∠*BHF* because they are alternate exterior angles:

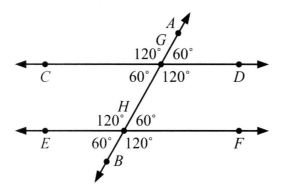

The answer is: The measure of ∠*BHF* is 120º.

---

example 2

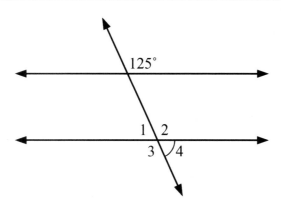

Lines *m* and *n* are parallel. What is the measure of ∠4?

Before determining the measure of any angle in the above figure, notice how the angles are labeled differently compared to the previous example. In this problem, the angles are labeled with numbers, while in the previous example each angle was labeled with three letters.

Using the method explained in **example 1,** being given the measure of one of the eight angles formed when a line crosses two parallel lines allows the other seven angles to be found:

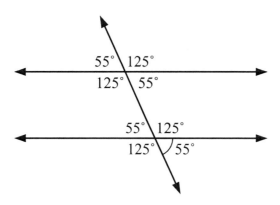

From the above diagram, it's easy to see that the measure of angle 4 is 55º.

---

example 3

In the figure shown below, $\overrightarrow{RS}$ is parallel to $\overrightarrow{TU}$, and $\overrightarrow{PT}$ intersects $\overrightarrow{RS}$ at Q.

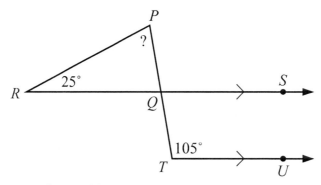

What is the measure of ∠RPQ?

The key to this problem is the fact that $\overrightarrow{RS}$ and $\overrightarrow{TU}$ are parallel. Even though they are rays, they can be treated as parallel lines. This means that $\overline{PT}$ is a transversal and that ∠T and ∠RQT are alternate interior angles. So if ∠T ≅ ∠RQT, then ∠RQT has a measure of 105º.

Next, you need to notice that ∠RQT and ∠RQP are supplementary angles, which means their combined measure totals 180º. If m∠RQT = 105º, then m∠RQP = 75º.

Lastly, using the angle sum property of triangles (explained in the next section), which states that the sum of the measures of the interior angles of a triangle is 180º, the measure of ∠RPQ can be found using the following equation:

$$m\angle QRP + m\angle RPQ + m\angle RQP = 180º$$

$$25º + m\angle RPQ + 75º = 180º$$

$$m\angle RPQ + 100º = 180º$$
$$ -100º \; -100º$$

$$m\angle RPQ = 80º$$

Practice Problems

Name _____

In the figure, $m \parallel n$. Use the figure to answer questions 1 – 10.

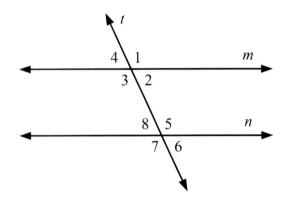

If angle 1 has a measure of 120°, what is the measure of:

1) ∠2

2) ∠3

3) ∠4

4) ∠5

5) ∠6

6) ∠7

7) Which angles are supplementary to ∠2?

8) Which angles are supplementary to ∠7?

9) If m∠1 = 120°, what is the sum of the measures of ∠2 and ∠4?

10) If m∠1 = 120°, what is the sum of the measures of ∠5 and ∠7?

THIS PAGE INTENTIONALLY BLANK

Lesson 34: Angle Sum Property of Triangles

Angle Sum Property of Triangles

The sum of the measures of the interior angles of a triangle is 180°. This is called the **angle sum property of triangles**, and it applies to ALL triangles, regardless of shape or size.

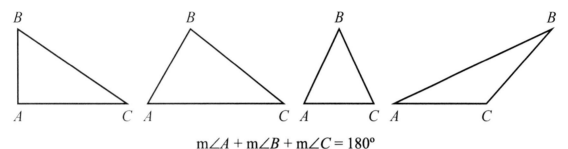

$$m\angle A + m\angle B + m\angle C = 180°$$

example 1

Use the angle sum property of triangles to find the measure of the missing angle.

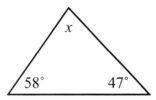

Since the sum of the measures of all three angles is 180°:

$$58° + 47° + x = 180°$$

$$105° + x = 180°$$
$$-105° \qquad -105°$$

$$x = 75°$$

Congruent angles were defined in the last section, but the sides of a **polygon** (a closed figure formed by line segments), such as a triangle, can be congruent as well. When a triangle has two or three congruent sides, it is called an isosceles triangle (if two sides are congruent) or an equilateral triangle (if all three sides are congruent).

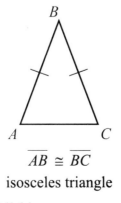

$\overline{AB} \cong \overline{BC}$

isosceles triangle

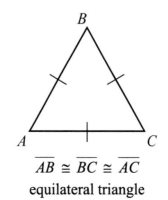

$\overline{AB} \cong \overline{BC} \cong \overline{AC}$

equilateral triangle

(Note: In a geometric diagram, drawing a little line through congruent sides of a polygon is a way of labeling those sides as congruent.)

One thing to note here, when a triangle has sides that are congruent, the angles opposite those sides are also congruent.

In the isosceles triangle, $\overline{AB} \cong \overline{BC}$, so $\angle A \cong \angle C$.

In the equilateral triangle, $\overline{AB} \cong \overline{BC} \cong \overline{AC}$, so $\angle A \cong \angle B \cong \angle C$, or in other words, all three angles of an equilateral triangle have the same measure, which happens to be 60°.

example 2

What is the measure of $\angle B$ in the figure below?

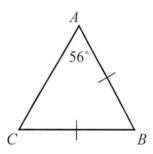

A. 34°

B. 56°

C. 62°

D. 68°

Sides $\overline{BC}$ and $\overline{AB}$ are congruent, so $\angle A$ and $\angle C$ are congruent.

Since the measures of all three angles must add up to 180°, the measure of $\angle B$ can be found adding together m$\angle A$ and m$\angle C$ and subtracting that from 180°:

$$m\angle A + m\angle B + m\angle C = 180°$$

$$56° + m\angle B + 56° = 180°$$

$$m\angle B + 112° = 180°$$
$$-112° \quad -112°$$

$$m\angle B = 68°$$

The answer is D.

example 3

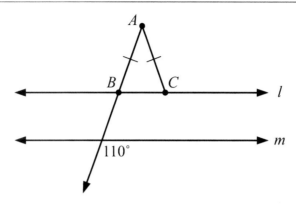

In the figure shown above, lines *l* and *m* are parallel, and △*ABC* is isosceles. What is the measure of ∠*ABC*?

A. 40º

B. 50º

C. 60º

D. 70º

This problem is asking for the measure of an interior angle of a triangle without providing any of the other two measures. However, even though this is an isosceles triangle like in **example 2**, it is the two parallel lines in the diagram will help us find our answer.

In Lesson 33, angles formed by transversals of parallel lines were explained. In this problem, eight angles are formed when $\overrightarrow{AB}$ crosses lines *l* and *m*, which are parallel. One of the eight angles formed is labeled as having a measure of 110º, and that is all we need to determine the measure of the other seven angles, one of which is ∠*ABC*. You should easily be able to label the measure of all of these angles as shown below:

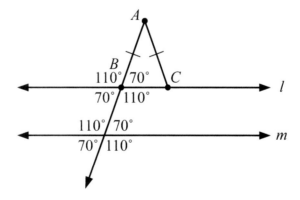

The measure of ∠*ABC* is 70º.

The answer is D.

Inscribed Triangles

An inscribed triangle is one whose **vertices** (plural of **vertex**) all lie on a figure, usually a circle.

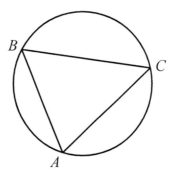

Points A, B, and C are the vertices of triangle ABC (also written $\triangle ABC$), and also lie on the circle.

A **diameter** of a circle is a line segment whose endpoints lie on a circle <u>and</u> which passes through the center of the circle. (*Any* line segment whose endpoints lie on a circle is called a **chord**.)

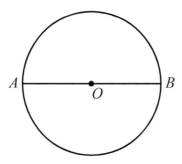

$\overline{AB}$ is a diameter of the circle because its endpoints lie on the circle and it passes through the center of the circle, point O.

Problems dealing with inscribed triangles often involve triangles that have one side that is also a diameter of the circle inscribing the triangle.

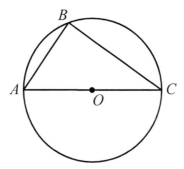

$\triangle ABC$ is inscribed in the circle, and $\overline{AC}$ is a diameter of the circle.

One important thing to notice about a triangle inscribed in a circle in this way: In the previous figure, if a line segment is drawn from the center (point *O*) to point *B*, two isosceles triangles are formed ($\triangle ABO$ and $\triangle BOC$).

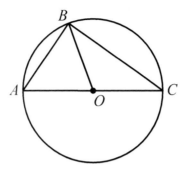

Any line segment drawn from the center of a circle to any point on the circle is called a **radius**.

In the figure below, $\overline{AO}$, $\overline{OC}$, and $\overline{OB}$ are **radii** (plural for radius) of the circle, and they are all congruent.

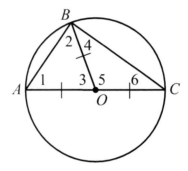

When sides of a triangle are congruent, the angles opposite those sides are congruent: $\angle 1 \cong \angle 2$, and $\angle 4 \cong \angle 6$

---

example 4

$\triangle ABC$ is inscribed in circle *O* and m$\angle A = 65°$

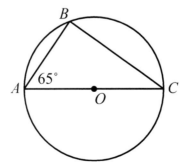

What is the measure of angle *C*?

Draw a line segment from point O to point B, creating two isosceles triangles out of the inscribed triangle.

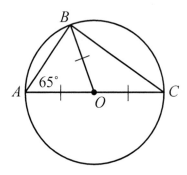

If m∠A = 65°, then m∠ABO must be 65° as well.

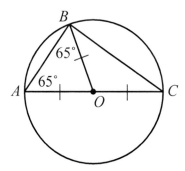

Using the angle sum property of triangles, the measure of the third angle in △ABO must be 50°.

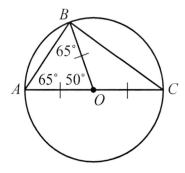

∠AOB and ∠BOC are supplementary angles, so if m∠AOB = 50°, then m∠BOC = 130°. That leaves 50° for the other two angles of △BOC. Since these remaining angles are congruent, they both have the same measure and must share the remaining 50° equally, meaning they each have a measure of 25°.

Answer: m∠C = 25°

Practice Problems

Name _____

The measures of two of the interior angles of a triangle are given. Find the measure of the third angle.

1) 50°, 60°

2) 30°, 60°

3) 20°, 100°

4) 45°, 90°

5) 45°, 45°

6) 35°, 55°

7) 10°, 130°

8) 64°, 115°

9) $x°$, 60°

10) $a°$, $(a + 10)°$

Use the figure below to find the measure of each angle.

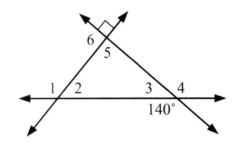

11) ∠1

12) ∠2

13) ∠3

14) ∠4

15) ∠5

16) ∠6

The degree measures of the interior angles of a triangle are given by the following expressions. Find the measures of the three angles.

17) $x°, (2x + 15)°, (2x – 10)°$

18) $2x°, (x + 30)°, (4x + 10)°$

19) $3x°, (2x + 10)°, (x – 10)°$

20) $4x°, (3x + 15)°, (5x – 15)°$

Use the figure below to answer questions 21–24.

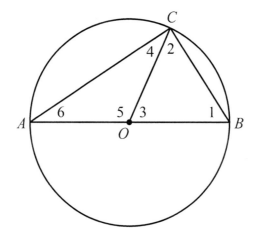

Given that m∠1 = 40°, find the measure of the following angles.

21) ∠6

22) ∠4

23) ∠2

24) ∠3

Lesson 35: The Pythagorean Theorem

<u>The Pythagorean Theorem</u>

One special kind of triangle is a **right triangle** — a triangle with one interior angle of 90°.

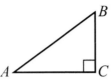

Note: In a polygon (like a triangle), the vertices (and corresponding angles) are labeled with capital letters, while the sides opposite each angle are labeled with lowercase letters.

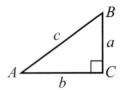

In a right triangle, the longest side (the side opposite the 90° angle) is called the **hypotenuse**, while the other two sides of a right triangle are called the **legs**.

One special relationship between the lengths of the sides of a right triangle is called the **Pythagorean Theorem**. It states that the sum of the squares of the measures of the legs equals the square of the measure of the hypotenuse.

For a triangle labeled with a and b as the legs and c as the hypotenuse, the Pythagorean Theorem can be expressed:

$$a^2 + b^2 = c^2 \qquad \textbf{Pythagorean Theorem}$$

This means that if the measures of two of the sides of a right triangle are known, the measure of the third side can be found.

---

example 1

The roads connecting the three towns on the map below form a right triangle. Two of the distances are given.

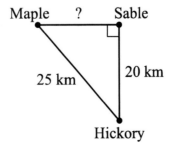

Based on the distances given on the map, what is the distance between Maple and Sable?

A. 12 km

B. 15 km

C. 16 km

D. 19 km

In any problem dealing with a right triangle, always be sure to identify the *hypotenuse* (the longest side). In this problem, the longest side is the one between Maple and Hickory and has a length of 25 kilometers. This will be used as c in the Pythagorean Theorem, while the other given distance can be used as either a or b:

$$a^2 + b^2 = c^2$$

$$20^2 + b^2 = 25^2$$

$$400 + b^2 = 625$$
$$-400 \qquad -400$$

$$b^2 = 225$$

$$\sqrt{b^2} = \sqrt{225}$$

$$b = 15$$

The answer is B.

---

example 2

The diagonal of a square is 25 units long. Which is the approximate length of a side of the square?

A. 18 units

B. 15 units

C. 5 units

D. 13 units

A **diagonal** is a line segment that connect vertices of a polygon through the interior of the polygon, so in the case of a square, it's a line segment that runs diagonally through the square from one corner to the opposite corner:

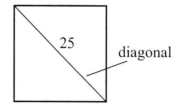

As can be seen in the previous diagram, a diagonal of a square divides that square into two right triangles, with the diagonal being the hypotenuse of both triangles:

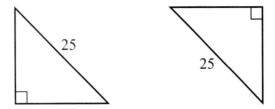

Because both right triangles came from a *square*, the legs of each triangle have the same length:

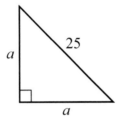

Now we can use the Pythagorean Theorem to find the approximate length of *a*:

$$a^2 + b^2 = c^2$$

$$a^2 + a^2 = 25^2$$

$$2a^2 = 625$$

$$\frac{2a^2}{2} = \frac{625}{2}$$

$$a^2 = 312.5$$

$$\sqrt{a^2} = \sqrt{312.5}$$

$$a \approx 17.7$$

The answer is A.

---

example 3

The lengths of the legs of a right triangle are 5 cm and 10 cm. Which of the following measures is closest to the length of the hypotenuse?

A. 11.2 cm

B. 11.4 cm

C. 11.6 cm

D. 11.8 cm

Use the Pythagorean Theorem to solve this problem:

$$a^2 + b^2 = c^2$$

$$(5)^2 + (10)^2 = c^2$$

$$25 + 100 = c^2$$

$$125 = c^2$$

$$\sqrt{125} = \sqrt{c^2}$$

$$c \approx 11.18$$

The answer is A.

Proving a Triangle to be a Right Triangle

The Pythagorean Theorem can also be used to prove that a triangle is a right triangle.

example 4

Is a triangle whose sides have lengths of 5, 12, and 13 a right triangle?

To answer this question, first identify the length of the longest side: 13 in this example. Then use this value as c, the length of the hypotenuse, and substitute the other 2 values as a and b (it doesn't matter which one is a or b).

$$a^2 + b^2 = c^2$$

$$5^2 + 12^2 \stackrel{?}{=} 13^2$$

$$25 + 144 \stackrel{?}{=} 169$$

$$169 = 169$$

These three lengths <u>do</u> make a right triangle.

example 5

Is a triangle whose sides have lengths of 6, 7, and 9 a right triangle?

$$a^2 + b^2 = c^2$$

$$6^2 + 7^2 \stackrel{?}{=} 9^2$$

$$36 + 49 \stackrel{?}{=} 81$$

$$85 \neq 81$$

This is <u>not</u> a right triangle.

example 6

A triangle is a right triangle if the lengths a, b, and c, of its three sides satisfy the following equation:

$$a^2 + b^2 = c^2$$

Which of the following is a right triangle?

A. a triangle with sides measuring 13, 15, and 19
B. a triangle with sides measuring 17, 25, and 36
C. a triangle with sides measuring 20, 21, and 29
D. a triangle with sides measuring 18, 23, and 41

To answer this question, just check the measures given in each answer and see if they work in the Pythagorean Theorem. The longest measure given must be used as c, but it doesn't matter which of the other two measures are chosen to be a and b:

Answer A: $a^2 + b^2 = c^2$

$13^2 + 15^2 \overset{?}{=} 19^2$

$169 + 225 \overset{?}{=} 361$

$394 \neq 361$

Answer B: $a^2 + b^2 = c^2$

$17^2 + 25^2 \overset{?}{=} 36^2$

$289 + 625 \overset{?}{=} 1296$

$914 \neq 1296$

Answer C: $a^2 + b^2 = c^2$

$20^2 + 21^2 \overset{?}{=} 29^2$

$400 + 441 \overset{?}{=} 841$

$841 = 841$

Answer D: $a^2 + b^2 = c^2$

$18^2 + 23^2 \overset{?}{=} 41^2$

$324 + 529 \overset{?}{=} 1681$

$852 \neq 1681$

The answer is C.

Other Triangle Problems

One thing to understand about *any* triangle, not just right triangles, is just common sense about how long the third side of a triangle could be given the measure of the other two sides.

For example, the longest side of a triangle (and remember that the hypotenuse is the longest side of a *right triangle only*) can't be longer than the sum of the lengths of the other two sides.

---

example 7

Eva has four sets of straws. The measurements of the straws are given below. Which set of straws could **not** be used to form a triangle?

A. Set 1: 4 cm, 4 cm, 7 cm
B. Set 2: 2 cm, 3 cm, 8 cm
C. Set 3: 3 cm, 4 cm, 5 cm
D. Set 4: 5 cm, 12 cm, 13 cm

For any triangle, the longest side can't be longer than the other two sides combined. To find which answer can **not** be a triangle, find the longest measure of each set of straws and see if it is longer than the sum of the lengths of the other two straws:

Answer A: The longest side is 7 cm, which is less than the sum of the other 2 sides (4 cm + 4 cm = 8 cm), so this set of straws *could* make a triangle.

Answer B: The longest side is 8 cm, which is <u>more</u> than the sum of the other 2 sides (2 cm + 3 cm = 5 cm), so this set of straws *could not* make a triangle.

Answer C: The longest side is 5 cm, which is less than the sum of the other 2 sides (3 cm + 4 cm = 7 cm), so this set of straws *could* make a triangle.

Answer D: The longest side is 13 cm, which is less than the sum of the other 2 sides (5 cm + 12 cm = 17 cm), so this set of straws *could* make a triangle.

The answer is B.

(Note: Even though the correct answer was found after checking Answers A and B, it is a good idea to keep going and check all four answers just in case you made a mistake.)

Trigonometry

Within a right triangle, there are three ratios involving the measures of two sides at a time that can be used for solving problems.

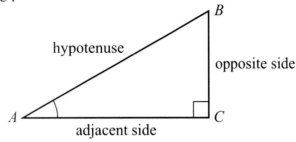

In terms of $\angle A$, the three sides of the right triangle can be labeled with reference to where that side is located relative to $\angle A$. There is one side opposite $\angle A$, a side next to (or adjacent to) $\angle A$, and the hypotenuse.

The three ratios are:

$$\text{the sine of } \angle A = \frac{\text{the measure of the side opposite } \angle A}{\text{the measure of the hypotenuse}}$$

$$\text{the cosine of } \angle A = \frac{\text{the measure of the side adjacent to } \angle A}{\text{the measure of the hypotenuse}}$$

$$\text{the tangent of } \angle A = \frac{\text{the measure of the side opposite } \angle A}{\text{the measure of the side adjacent to } \angle A}$$

These are usually abbreviated as:

$$\sin A = \frac{\text{opp.}}{\text{hyp.}} \qquad \cos A = \frac{\text{adj.}}{\text{hyp.}} \qquad \tan A = \frac{\text{opp.}}{\text{adj.}}$$

These ratios are called trigonometric functions (or trig. functions for short).

---

example 8

Find the tan A, sin A, and cos A of the given triangle.

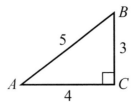

$$\sin A = \frac{\text{opp.}}{\text{hyp.}} \qquad \cos A = \frac{\text{adj.}}{\text{hyp.}} \qquad \tan A = \frac{\text{opp.}}{\text{adj.}}$$

$$\sin A = \frac{3}{5} \qquad \cos A = \frac{4}{5} \qquad \tan A = \frac{3}{4}$$

On your calculator, either on a button or just above a button, you can see $\sin^{-1}$, $\cos^{-1}$, and $\tan^{-1}$.

These are inverse functions, which will swap the angle and the value of the trig. function, as in the following example.

From above, $\sin A = \dfrac{3}{5}$, so $\sin^{-1}\left(\dfrac{3}{5}\right) = A$

Using the inverse function on your calculator, you can find the measure of the angle itself:

$$\text{if } A = \sin^{-1}\left(\frac{3}{5}\right)$$

$$\text{then } A \approx 37°$$

Note: When using these trig. functions, make sure your calculator is in "degree" mode and not "radian" mode.

Practice Problems

Name _____

If c is the measure of the hypotenuse of a right triangle and a and b are the legs of the triangle, find each missing measure.

1) $a = 3, b = 4, c =$ _____

2) $a = 5, b =$ _____ $, c = 13$

3) $a =$ _____ $, b = 1, c = \sqrt{2}$

4) $a = 2, b = 3, c =$ _____

Determine whether the following lengths would form the sides of a right triangle.

5) $9, \sqrt{40}, 11$

6) $\sqrt{13}, 8, 9$

7) $\sqrt{3}, 2, \sqrt{6}$

8) $20, 21, 29$

Determine whether the given lengths could be the three sides of *any* triangle.

9) $5, 6, 9$

10) $11, 14, 26$

11) $11, 12, 23$

12) $9, 15, 22$

Use the figure below to answer question 13.

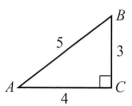

13) Find sin *A*, cos *A*, and tan *A*.

Use your calculator to solve question 14–16. (Make sure your calculator is in *degree* mode.)

14) If sin *M* = 0.7660, what is the measure of ∠*M*?

15) If cos *N* = 0.3420, what is the measure of ∠*N*?

16) If tan *P* = 1.804, what is the measure of ∠*P*?

For each triangle below, find the measure of the marked angle to the nearest degree.

17)   18)

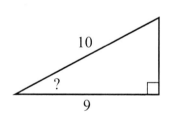   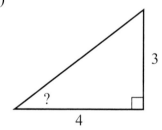

For each triangle below, find the measure of the marked side.

19)   20)

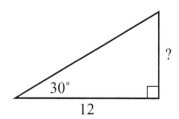   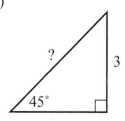

Lesson 36: Interior Angles, Exterior Angles, and Diagonals of Polygons

Interior Angles of Polygons

The angle sum property of triangles can help determine the sum of the measures of interior angles of other polygons.

The sum of the measures of the interior angles of a triangle is 180°, so to find the sum of the measures of the interior angles of a polygon of more than three sides, see how many triangles into which the polygon can be divided with the vertices of the polygon also being the vertices of the triangles.

---

example 1

Find the sum of the measures of the interior angles of the given hexagon.

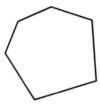

The line segments to be drawn in order to divide up the hexagon into triangles are called *diagonals*.

By drawing three diagonals (that don't intersect), the hexagon is divided into four triangles, each with an interior angle sum of 180°. So the sum of the measures of the interior angles of the hexagon is $4 \times 180° = 720°$.

The following table summarizes the sum of the measures of the interior angles of other common polygons:

polygon	number of sides	number of triangles formed by diagonals	sum of the measures of the interior angles
quadrilateral	4	2	$2 \times 180° = 360°$
pentagon	5	3	$3 \times 180° = 540°$
octagon	8	6	$6 \times 180° = 1080°$
n-gon	n	$n - 2$	$(n - 2) \times 180°$

example 2

Harry measured all but one angle of a hexagon. The total degree measure for all of the angles he measured was 550°. What is the measure, in degrees, of the remaining angle?

A. 92°

B. 120°

C. 170°

D. 720°

There are two ways to approach this problem.

One method is to use the formula for the sum of the measures of the interior angles of a polygon, which is $(n - 2) \times 180°$, and subtract 550° from the total to see what the measure of the remaining angle should be. Since there are 6 sides in a hexagon, n will equal 6:

$$\text{sum of the measures of interior angles} = (n - 2) \times 180°$$
$$= (6 - 2) \times 180°$$
$$= 4 \times 180°$$
$$= 720°$$

Now subtract 550° from 720° to see what the measure of the sixth interior angle is:

$$720° - 550° = 170°$$

The answer is C.

The second method, if you can't remember the formula for the sum of the measures of the interior angles of a polygon, is to draw a sketch of the polygon and divide it into as many non-overlapping triangles as you can:

This was shown in **example 1**, and you can do this to see how many triangles you can get, as long as you remember that the sum of the measures of the interior angles of a triangle is 180°. This will also result in a sum of 720°, which you can use to find the measure of the missing angle.

Exterior Angles of Polygons

An **exterior angle** of a polygon is formed by extending one side of the polygon.

The sum of the measures of the exterior angles of a polygon is *always* 360º, no matter how many sides the polygon has or whether or not it is a regular polygon. (A **regular polygon** is a polygon whose sides are all congruent and whose interior angles are all congruent.)

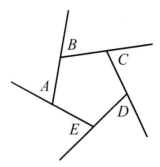

$$m\angle A + m\angle B + m\angle C + m\angle D + m\angle E = 360º$$

This is true no matter how many sides a polygon has, as shown by the square and octagon below:

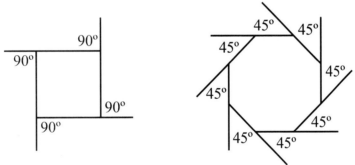

The sum of the measures of the exterior angles of a polygon with *n* sides equals 360º. This means that the exterior angles of a *regular* polygon with *n* sides each has a measure of 360º divided by *n*.

The most important aspect of an exterior angle, however, is that when it is paired up with the adjacent interior angle, the two angles make a linear pair. You should remember from Lesson 32, a linear pair is made up of two supplementary angles whose non-common sides make a line:

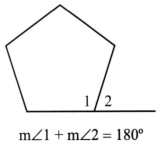

$$m\angle 1 + m\angle 2 = 180º$$

Diagonals of Polygons

Any polygon can have more diagonals than just the ones that divide a polygon into triangles.

Any quadrilateral (four-sided polygon) has two diagonals.

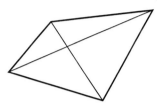

Any pentagon has five diagonals.

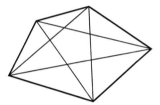

Any hexagon has nine diagonals.

example 3

How many diagonals will an octagon have?

There is a pattern to how many diagonals are added each time a side is added to a polygon:

number of sides	3	4	5	6
number of diagonals	0	2	5	9

↘+2↗   ↘+3↗   ↘+4↗

A polygon with seven sides would have 9 + 5, or 14, diagonals, so an octagon would have 14 + 6, or 20 diagonals.

Practice Problems

Name _____

For problems 1–6, find the sum of the measures of the interior angles of the polygon with the given number of sides.

1) 5

2) 6

3) 8

4) 21

5) 4

6) 7

For problems 7–10, determine the number of sides a regular polygon would have with the given value for each interior angle. [<u>Hint</u>: $(n-2) \times 180° =$ size of interior angle $\times n$]

7) 144°

8) 120°

9) 135°

10) 108°

11) The measures of four interior angles of a pentagon are 120°, 95°, 100°, and 115°. Find the measure of the fifth interior angle.

12) Five interior angles of a hexagon each have a measure of 125°. Find the measure of the sixth interior angle.

For problems 13–18, determine the measure of an exterior angle of a regular polygon with the given number of sides. (Remember, the sum of the measures of *all* exterior angles of a polygon is 360°.)

13) 5   14) 3

15) 8   16) 6

17) 4   18) 9

For problems 19–24, state the number of diagonals *from each vertex* that each polygon with the given number of sides has.

19) 5   20) 12

21) 15   22) 8

23) 7   24) 10

For problems 25–30, state the *total number of diagonals* that each polygon with the given number of sides has.

25) 8   26) 10

27) 11   28) 9

29) 14   30) 7

Lesson 37: Midpoint and Distance Formulas

<u>Midpoint Formula</u>

To find the number exactly between two other numbers, average together those two numbers.

On a number line, the number exactly between two other numbers is the **midpoint** between those other two positions on the number line.

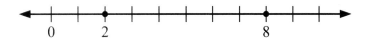

To find the midpoint between 2 and 8, average them together:

$$\frac{2+8}{2} = \frac{10}{2} = 5$$

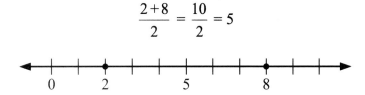

If we labeled these points, A, B, and M (for midpoint), $\overline{AM} \cong \overline{MB}$.

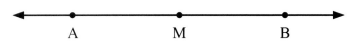

Two number lines that intersect perpendicularly make a coordinate plane where a point has an x- and y-value to define its location.

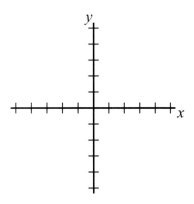

To find the midpoint of two points plotted on a coordinate plane, average the two x-values of the points for the x-value of the midpoint, and average the two y-values of the points for the y-value of the midpoint.

$$\text{midpoint} = \left(\frac{x_1 + x_2}{2}, \frac{y_1 + y_2}{2} \right)$$

example 1

Find the midpoint of the line segment with endpoints of $(-3, 1)$ and $(5, -7)$.

Let $x_1 = -3$ and $y_1 = 1$, and let $x_2 = 5$ and $y_2 = -7$.

$$\text{midpoint} = \left(\frac{x_1 + x_2}{2}, \frac{y_1 + y_2}{2} \right)$$

$$= \left(\frac{-3 + 5}{2}, \frac{1 + (-7)}{2} \right)$$

$$= \left(\frac{2}{2}, \frac{-6}{2} \right)$$

$$= (1, -3)$$

Distance Formula

How far apart are the two points plotted on the coordinate plane?

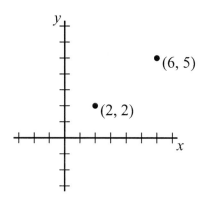

The Pythagorean Theorem can be used to solve this problem if a right triangle is sketched onto the graph.

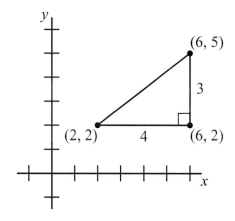

The lengths of the legs are 3 and 4, so the length of the hypotenuse, which is the distance between these two points, is 5.

$$a^2 + b^2 = c^2$$

$$4^2 + 3^2 = c^2$$

$$16 + 9 = c^2$$

$$25 = c^2$$

$$c = 5$$

For any two points on a graph, a more generic form of the Pythagorean Theorem can be produced.

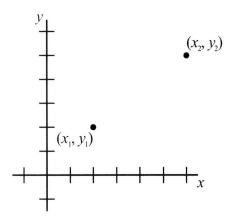

Again sketch a right triangle onto the graph.

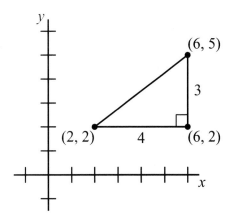

The length of one leg is the second x-value minus the first x-value ($x_2 - x_1$), and the length of the other leg is the second y-value minus the first y-value ($y_2 - y_1$).

Substitute these expressions into the Pythagorean Theorem for a and b:

$$c^2 = a^2 + b^2$$

$$c^2 = (x_2 - x_1)^2 + (y_2 - y_1)^2$$

Replace c with d (for distance):

$$d^2 = (x_2 - x_1)^2 + (y_2 - y_1)^2$$

Then take the square root of both sides:

$$d = \sqrt{(x_2 - x_1)^2 + (y_2 - y_1)^2}$$ and this is the **distance formula**.

---

example 2

Find the distance between $(3, 2)$ and $(-6, 0)$.

Let $x_1 = 3$, $y_1 = 2$, $x_2 = -6$, and $y_2 = 0$, and substitute these values into the distance formula.

$$d = \sqrt{(x_2 - x_1)^2 + (y_2 - y_1)^2}$$

$$d = \sqrt{(-6 - 3)^2 + (0 - 2)^2}$$

$$d = \sqrt{(-9)^2 + (-2)^2}$$

$$d = \sqrt{81 + 4}$$

$$d = \sqrt{85}$$

The distance between the given points is $\sqrt{85}$, or about 9.2.

Practice Problems

Name _____

Find the coordinates of the midpoint of line segment AB.

1) $A(6, 4), B(2, 2)$

2) $A(-4, 3), B(2, -5)$

3) $A(5, -1), B(-5, 1)$

4) $A(4, 7), B(-4, 1)$

5) $A(-6, 2), B(-4, -2)$

6) $A(-5, -6), B(-1, -8)$

7) $A(1, -8), B(-2, 3)$

8) $A(6, 4), B(9, -5)$

If M is the midpoint of line segment AB, find the coordinates of the missing endpoint.

9) $A(-4, -1), M(-1, 2)$

10) $B(5, 6), M(5, 1)$

11) $B(-3, 7), M(0, 0)$

12) $A(-8, -6), M(-4, -6)$

Find the distance between the points with the given coordinates. Leave answers in simplest radical form.

13) $(0, -4), (3, 0)$ 14) $(4, 7), (4, -2)$

15) $(-1, 3), (7, 9)$ 16) $(-7, 1), (5, 6)$

17) $(-4, -2), (1, -2)$ 18) $(-2, 6), (2, 3)$

19) $(-5, 0), (0, 4)$ 20) $(2, 1), (7, 4)$

Find the two possible values of p if the points with the given coordinates have the indicated distance apart.

21) $(-5, 4), (p, 1); d = 5$ 22) $(2, p), (-4, 3); d = \sqrt{52}$

Lesson 38: Ratios and Proportions with Congruent and Similar Polygons

Congruent Polygons

Congruent polygons are polygons whose corresponding sides are congruent and whose corresponding angles are congruent. (Remember, corresponding sides or angles of two or more polygons are those sides or angles that are in the same relative position.) In other words, the congruent polygons look exactly alike.

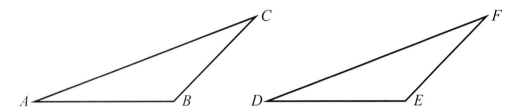

In the above figure, $\triangle ABC \cong \triangle DEF$ if $\overline{AB} \cong \overline{DE}$, $\overline{BC} \cong \overline{EF}$, $\overline{AC} \cong \overline{DF}$, and if $\angle A \cong \angle D$, $\angle B \cong \angle E$, $\angle C \cong \angle F$.

The notation used to show this on a **diagram** looks like this:

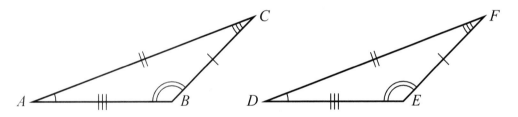

Also, two congruent polygons can be oriented in any direction:

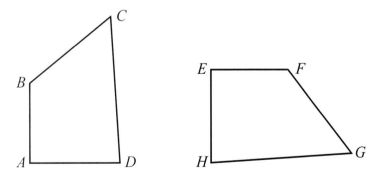

In many geometry courses, there will be problems that involve *proving* that one polygon is congruent to another. To understand the basics, as long as you know what congruent polygons *are*, you should be fine. Just remember that *congruent polygons look exactly alike*, because corresponding sides are congruent and corresponding angles are congruent.

example 1

Triangles *ABC* and *DEF* shown below are congruent.

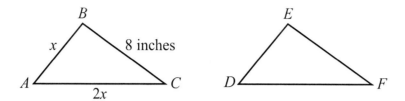

The perimeter of △*ABC* is 23 inches. What is the length of side $\overline{DF}$ in △*DEF*?

The sides of congruent figures are also congruent (have the same measure). Since the length of $\overline{BC}$ is 8 inches, the length of $\overline{EF}$ is also 8 inches. We know this without doing any work. So, in order to find the length of side $\overline{DF}$, we need to find the length of side $\overline{AC}$.

The key to finding the length of side $\overline{AC}$ is figuring out what *x* equals, and this can be found because we know the perimeter, which is the sum of the measures of all three sides of the triangle:

$$x + 2x + 8 = 23$$

Solving this equation for x:

$$x + 2x + 8 = 23$$
$$3x + 8 = 23$$
$$-8 \quad -8$$
$$3x = 15$$
$$\frac{3x}{3} = \frac{15}{3}$$
$$x = 5$$

Since $\overline{AC}$ has a length of 2*x*, or 2 × 5, $\overline{DF}$ has the same measure.

The length of $\overline{DF}$ is 10 inches.

Similar Polygons

Similar polygons are polygons whose corresponding angles are congruent, but *not* their corresponding sides. In other words, a polygon that is similar to another polygon has the same exact shape, but is bigger or smaller than the first polygon.

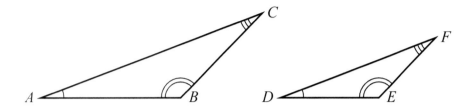

In the above figure, $\triangle ABC \sim \triangle DEF$ (read, "triangle ABC is similar to triangle DEF").

While corresponding sides of similar polygons are not congruent, corresponding sides are *proportional* to each other.

In the figure above, the ratio of the measure of $\overline{AB}$ to the measure of $\overline{DE}$ is equal to the ratio of the measure of $\overline{BC}$ to the measure of $\overline{EF}$, and is also equal to the ratio of the measure of $\overline{AC}$ to the measure of $\overline{DF}$. (Note: A line segment is labeled by its two endpoints, with a small line drawn over them. However, the measure of a line segment is labeled with the same two endpoints, but without the small line over the letters.)

This can be expressed as:

$$\frac{AB}{DE} = \frac{BC}{EF} = \frac{AC}{DF}$$

This relationship will be further explored in the example on the next page.

As previously mentioned, similar polygons have congruent corresponding angles, but not congruent corresponding sides (or else they would be congruent polygons).

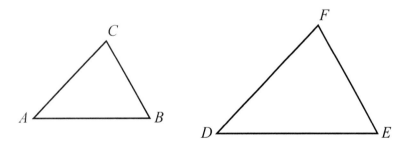

$\triangle ABC \sim \triangle DEF$ if $\angle A \cong \angle D$, $\angle B \cong \angle E$, and $\angle C \cong \angle F$

Again, this means that

$$\frac{AB}{DE} = \frac{BC}{EF} = \frac{AC}{DF}$$

When setting up this proportion, be careful to keep all the information from one polygon in the numerators and all the information from the other polygon in the denominators.

Such a proportion can be used to find the lengths of the sides of one of the similar polygons, as long as *one* of the lengths from that polygon is already known.

---

example 2

In the diagram below, $\triangle ABC$ and $\triangle DEF$ are similar triangles with the dimensions shown, in units.

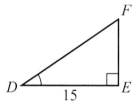

What is the length, in units, of $\overline{EF}$?

Since $\dfrac{BC}{EF} = \dfrac{AC}{DF}$, a proportion can be set up to be solved for *EF*.

$BC = 6$, $DF = 15$, and $AC = 9$

$$\frac{BC}{EF} = \frac{AC}{DF}$$

$$\frac{6}{EF} = \frac{9}{15}$$

cross-multiply:

$$6 \cdot 15 = 9 \cdot EF$$

and solve for the missing value:

$$90 = 9 \cdot EF$$

$$\frac{90}{9} = \frac{9 \cdot EF}{9}$$

$$EF = 10$$

The length of $\overline{EF}$ is 10.

Practice Problems

Name _____

Given that $ABCD \cong JKLM$, identify the corresponding sides and angles.

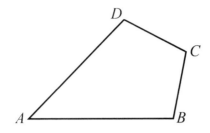
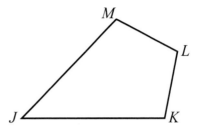

1) $\overline{AB} \cong$ _____   2) $\angle L \cong$ _____

3) $\angle B \cong$ _____   4) $\overline{LM} \cong$ _____

Given that $ABCDE \cong JKLMN$, identify the corresponding sides and angles.

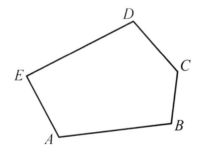
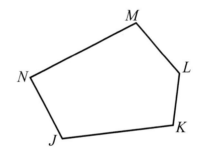

5) $\angle E \cong$ _____   6) $\overline{BC} \cong$ _____

7) $\overline{MN} \cong$ _____   8) $\angle A \cong$ _____

For the given measurements, determine if the triangles are *similar* or *not similar*.

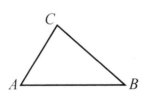
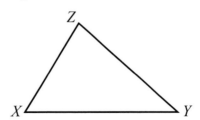

9) $AB = 4$, $BC = 6$,   10) $AC = 5$, $BC = 7$
    $XY = 8$, $YZ = 12$ $XZ = 8$, $YZ = 11$

Use the diagram below to answer questions 11–14.

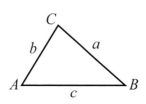
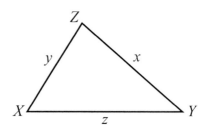

Triangle *ABC* is similar to triangle *XYZ*. Use the given the measures of the sides of each triangle to find each missing length.

11) $a = 5, b = 7, x = 15, y = ?$

12) $b = 6, c = 8, y = 9, z = ?$

13) $b = 2, c = 3, y = 5, z = ?$

14) $a = ?, b = 6, x = 6, y = 4$

15) A 6-foot tall man casts a 4-foot shadow at the same time that a nearby 54-foot tall tree casts a shadow. How long is the tree's shadow?

16) A 5-foot tall girl is standing 3 feet from a mirror placed on the ground. She can see the top of a flagpole reflected in the mirror. If the mirror is 18 feet from the base of the flagpole, how tall is the flagpole?

Lesson 39: Transformations

A **transformation** is a change in location, orientation, or size of a point, line, or polygon. The new figure created after a transformation is called an **image**.

There are four types of transformations that you could be asked to perform:

<u>Translations</u>

A **translation** is a transformation of a figure in which each point of the figure is moved the same distance in the same direction. (Translations are often shown on a coordinate plane so the distance moved can be easily measured.)

example 1

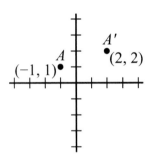

Point A is translated three units to the right and one unit up to point A'.

(Note: Whenever the transformation of a point occurs, the new point usually has the same label but with an apostrophe (') written next to it. A' is said, "A prime".)

example 2

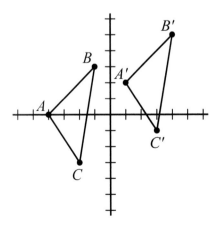

$\triangle ABC$ is translated to $\triangle A'B'C'$ by moving it five units to the right and two units up. $\triangle A'B'C'$ is the image of $\triangle ABC$.

Reflections

A **reflection** is a transformation that involves flipping a figure over a line called a **line of symmetry**. The line of symmetry acts like a mirror, with the original figure on one side of the line and the mirror image of that figure on the other side of the line.

---

example 3

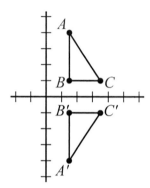

$\triangle A'B'C'$ is the image of $\triangle ABC$ after a reflection about the *x*-axis.

---

example 4

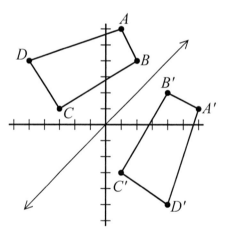

This graph shows the reflection of a quadrilateral about the line $y = x$.

Rotations

A **rotation** is a transformation that involves turning a figure around a point called the **point of rotation**. Figures can be rotated clockwise or counterclockwise, and the point of rotation can be a point within the figure itself (like a vertex point), or a point anywhere on the coordinate plane.

---

example 5

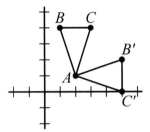

△*ABC* is rotated 90° clockwise about point *A*.

---

example 6

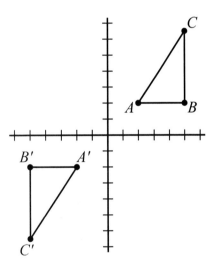

△*ABC* is rotated 180° about the origin. (Note: A 180° rotation is the same whether it's clockwise or counterclockwise.)

Dilations

A **dilation** is a transformation that produces an image that is the same shape as the original, but is a different size. The description of a dilation includes the **scale factor** and the **center of dilation**.

(Note: Most dilations in coordinate geometry use the origin (0, 0) as the center of the dilation.)

---

example 7

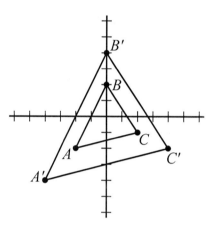

△*A'B'C'* is the dilation of △*ABC* with the center of dilation at the origin and a scale factor of 2.

---

example 8

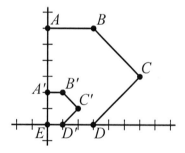

Pentagon *A'B'C'D'E'* is the image of Pentagon *ABCDE* with the center of dilation at the origin (which is also Point *E*) and a scale factor of $\frac{1}{3}$.

Practice Problems

Name _____

Sketch each transformation on the graph provided.

1) Translate Point *A* six units to the right and four units down.

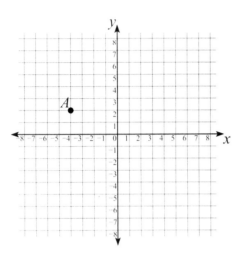

2) Translate △*ABC* seven units up and five units to the left.

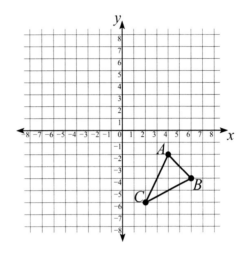

3) Reflect Figure *ABCD* about the *y*-axis.

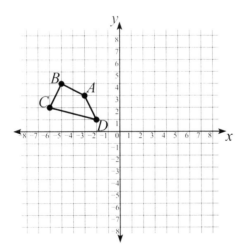

4) Reflect △*ABC* about the line $y = x$.

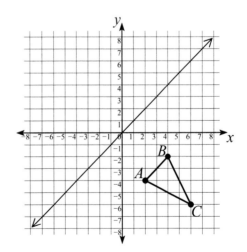

Sketch each transformation on the graph provided.

5) Rotate Point *A* clockwise 90°
    about the origin.

6) Rotate △*ABC* 180° about Point A.

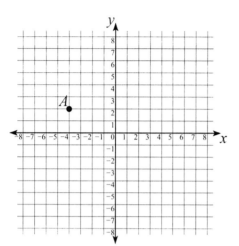

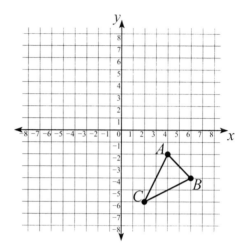

7) Dilate Figure *ABCD* with the center
    of dilation at the origin and a scale
    factor of 2.

8) Dilate △*ABC* with the center or
    dilation at the origin and a scale
    factor of ⅓.

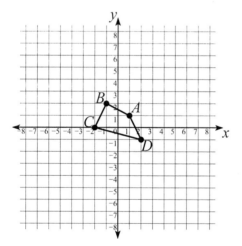

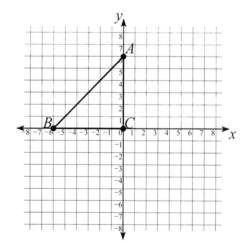

Lesson 40: Measurement

Measurement involves determining the ratio of the magnitude of a *quantity* to the magnitude of a *unit*. A measurement is a ratio because the quantity and unit can change, but the overall ratio will remain the same. For example, 6 feet is the same as 2 yards or 72 inches.

Common Measurements

There are five basic types of measurements: length, area, weight (or mass), volume, and time. In math, you generally will deal with four of these five.

1) *Length*

Length is a one-dimensional measurement, expressed in units raised to the first power, such as feet, meters, miles, inches, etc.

Length measures the distance between two points.

2) *Area*

Area is a two-dimensional measurement, expressed in units raised to the second power, such as square centimeters (sq. cm or cm^2), square feet (sq. ft or ft^2), etc.

Area measures two-dimensional space, or the size of a surface.

3) *Volume*

Volume is a three-dimensional measurement, expressed in units raised to the third power, such as cubic inches (cu. in or in^3), cubic meters (cu. m or m^3), etc.

Volume measures three-dimensional space, or capacity.

4) *Time*

Time is a one-dimensional measurement, expressed in units raised to the first power, such as seconds, minutes, hours, days, etc.

Time does not measure a physical dimension; it is how the period (or interval) between events is measured.

example 1

For which of the following would a yard be the **most** appropriate unit of measure?

A. the area of a street sign
B. the volume of a bus
C. the length of a cafeteria
D. the distance from Boston to Springfield, Massachusetts

The first thing to determine about this problem is the kind of measurement for which a "yard" would be used, such as whether it is for measuring a length, area, or volume.

An area could be measured in "square yards", and a volume could be measured in "cubic yards", but length (or distance) is just "yards". This eliminates Answers A and B.

To determine whether a yard would be appropriate for measuring the length of a room or the distance between two cities, you need to understand how big a yard is and how this compares to the two choices. A cafeteria could be 100 feet long, while Boston and Springfield are almost 100 miles apart. Even if you don't know the distance between these cites, you should still know that a *yard* is closer to a *foot* than it is to a *mile*.

The answer is C.

Rounding Off

To round off a number:

1) Find the place value to which you want to round off (the "rounding digit"), and look at the digit to its right.

2) If the digit to the right is less than five, do not change the "rounding digit", but change all digits to the right of it to zero (and any digits to the right of the decimal point get dropped).

3) If the digit to the right of the rounding digit is greater than or equal to five, add one to the rounding digit and change all the digits to its right to zero (and any digits to the right of the decimal point get dropped).

---

example 2

When Matt's and Damien's broad jumps were measured **accurately to the nearest foot**, each measurement was 21 feet. Which statement best describes the greatest possible difference in the lengths of Matt's jump and Damien's jump?

A. One jump could be up to $\dfrac{1}{4}$ foot longer than the other.

B. One jump could be up to $\dfrac{1}{2}$ foot longer than the other.

C. One jump could be up to 1 foot longer than the other.

D. One jump could be up to 2 feet longer than the other.

When rounding to a whole number, the original value could be half of a unit above or below the rounded value. For example, 21.49 would be rounded to 21, and 20.50 would also be rounded to 21. The difference between 21.49 and 20.50 is almost 1.

The answer is C.

Systems of Measurement

English Standard System (U.S. Customary System)

This system of measurement developed out of the way people measured for themselves distances (length), weights (mass), and capacities (volume).

People measured short distances with their feet, long distances were measured in paces (a "mile" was 1,000 paces), and capacities were measured with common household items such as cups, pails, and baskets.

Eventually each unit of measurement was standardized (which is helpful since each person's foot is a different size), but we still ended up with many different units of measurement. To measure distance (length), there is the inch, foot, yard, or mile. To measure weight (mass), there is the ounce, pound, or ton. And to measure capacity (volume), there is the teaspoon, tablespoon, cup, pint, quart, or gallon.

Having so many different units of measurement would sound confusing if you weren't already used to them, but it gets worse when you realize that there is no standard way of converting from one unit of measure to another. Each conversion is different:

Distance (Length)	Weight (Mass)	Capacity (Volume)
12 inches (in) = 1 foot (ft)	16 ounces (oz) = 1 pound (lb)	3 teaspoons (tsp) = 1 tablespoon (tbsp)
3 feet (ft) = 1 yard (yd)	2000 pounds (lbs) = 1 ton	16 tablespoons (tbsp) = 1 cup (c)
5280 feet = 1 mile (mi)		8 fluid ounces (fl oz) = 1 cup (c)
		2 cups = 1 pint (pt)
		2 pints = 1 quart (qt)
		4 quarts = 1 gallon (gal)

As you can see in the table above, there is no standard conversion between measurements. 12 inches are in 1 foot, but 16 ounces are in 1 pound, and so on. And there are *many* more conversions in the English Standard System that are not listed in the table above. Most of the time you need a reference sheet available whenever you need to make a conversion.

---

example 3

The Massachusetts Highway Department is responsible for more than 66.5 million feet of roadway in Massachusetts. **About** how many miles is 66.5 million feet?

A. 4,200 miles

B. 35,100 miles

C. 12,600 miles

D. 11,700 miles

To answer this question, you need to know the relationship between feet and miles. There are 5280 feet in 1 mile, so you just need to divide 66.5 million by 5280:

$$66,500,000 \div 5280 \approx 12,595$$

The answer is C.

(Note: While this looks easy enough, measurements include quantities *and* units. A method of measurement conversion called **dimensional analysis** will be covered in the next lesson.)

The Metric System

The metric system is a system of measurement based on the *meter, kilogram,* and *liter.* A **meter** (m) is the basic unit of length in the metric system, and is equal to about 3.28 feet. A **kilogram** (km) is the basic unit of mass in the metric system, and is equal to about 2.2 pounds. A **liter** (L) is the basic unit of volume in the metric system, and it slightly larger than a quart (about 1.05 quarts).

You may have noticed that one of these three units of measurement has the prefix "kilo-" while the other two do not. It is prefixes like this that makes the metric system so easy to use, and why all but several countries (the United States being one of these few) have switched to the metric system.

The reason is it so simple to use is the metric system has only one unit for any physical quantity. For example, with the use a prefix, the *meter* can used for any distance or length, no matter how big or small.

Prefix	Symbol	Scale	Decimal Equivalent	10^n
giga	G	billion	1,000,000,000	10^9
mega	M	million	1,000,000	10^6
kilo	k	thousand	1,000	10^3
hecto	h	hundred	100	10^2
deca	da	ten	10	10^1
(none)	(none)	one	1	10^0
deci	d	tenth	0.1	10^{-1}
centi	c	hundredth	0.01	10^{-2}
milli	m	thousandth	0.001	10^{-3}
micro	μ	millionth	0.000001	10^{-6}
nano	n	billionth	0.000000001	10^{-9}

As shown in the table above, the word *kilo*meter, means *one thousand* meters. The abbreviation for the kilometer is "km" — "k" for kilo and "m" for meter.

The metric system is a decimal-based system because it is based on multiples of ten. This means that any measurement given in one metric unit can be converted to another metric unit just by moving the decimal point.

example 4

Which of the following is equivalent to 20 centimeters?

A. 2000 millimeters

B. 200 millimeters

C. 20 millimeters

D. 2 millimeters

To answer this question, you need to know the relationship between centimeters and millimeters. The metric system makes measurement conversions easy because all metric units are separated by a factor of 10, 100, 1000, etc. All you need to do is move the location of the decimal point the appropriate number of places.

Since there are 10 millimeters in 1 centimeter, these two units of measure differ by a factor of 10. This means the decimal point only needs to move one place to make this conversion. But is it a move to the left or to the right?

When you are changing from a larger to a smaller unit of measure, move the decimal point to the *right*. (1 centimeter becomes 10 millimeters.) If you are changing from a smaller unit of measure to a larger one, move the decimal point to the *left*. (10 millimeters becomes 1 centimeter.)

In this problem, since we are changing from a larger unit of measure to a smaller one (centimeters to millimeters), move the decimal point to the right one place:

20 centimeters = 200 millimeters

The answer is B.

Conversions Between Measurement Systems

Common conversions between the English Standard System and the Metric System are shown in the table below:

Length	Mass	Volume
1 inch = 25.40 mm	1 pound = 453.6 grams	1 fl oz = 29.57 mL
1 foot = 30.48 cm	1 kilogram = 2.205 pounds	1 cubic inch = 16.39 mL
1 mile = 1.609 km		1 gallon = 3.785 L
1 meter = 39.37 in		1 cubic foot = 28.32 L
1 km = 0.621 mi		1 liter = 1.057 quarts

Practice Problems

Name _____

Round each number to the nearest *tens* place.

1) 87.4   2) 105   3) 1,010,101

Round each number to the nearest *tenths* place.

4) 6.52   5) 3.149   6) 17.1

Round each number to the nearest *thousands* place.

7) 988   8) 12,345   9) 9,031,500

Round each number to the nearest *ones* place.

10) 13.7   11) 0.94   12) $53\dfrac{4}{9}$

Calculate the following Standard conversions.

13) Convert 3 pounds into ounces.   14) Convert 18 quarts into gallons.

15) Convert 32 tablespoons into pints.   16) Convert 2,150 tons into pounds.

17) Convert 54 feet into yards.   18) Convert 2 miles into feet.

Calculate the following Metric conversions.

19) Convert 400 meters into kilometers.   20) Convert 3 kilograms into milligrams.

21) Convert 0.05 liters into milliliters.   22) Convert 0.0004 meters into nanometers.

23) Convert 1000 decigrams into decagrams.   24) Convert 340 milliliters into centiliters.

Calculate the following conversions between Standard and Metric.

25) Convert 18 inches into centimeters.   26) Convert 80 kilometers into miles.

27) Convert 10 gallons into liters.   28) Convert 6 feet into meters.

29) Convert 32 fluid ounces into milliliters.   30) Convert 55 kilograms into pounds.

Lesson 41: Dimensional Analysis

Dimensional Analysis

Many math problems involve answers with *labels*, such as "3 feet" instead of just "3", or "45 miles per hour" instead of just "45". (These labels are also called "units" or "units of measurement".)

Whenever you need to change the *labels* (or *units*) in a problem, you can use a process called **dimensional analysis**, which involves using ratios and setting up proportions.

Before getting into the steps involved with dimensional analysis, take a step back and recall a rule about fractions: If you need to change the numerator or denominator of a fraction, *both* must be multiplied (or divided) by the same number so that the overall value of the fraction is not changed.

For example, to add $\frac{2}{3} + \frac{7}{12}$, both fractions need a common denominator of 12. The first fraction can have its denominator changed to 12 by multiplying the numerator *and* denominator by 4:

$$\frac{2}{3} \times \frac{4}{4} = \frac{8}{12}$$

$\frac{4}{4}$ is a fraction called a **conversion factor**, which always has a value of 1. (Any fraction with the same value in the numerator *and* denominator is equal to 1.)

In this way, the value of the original fraction isn't changed since anything multiplied by 1 equals itself.

$$\frac{2}{3} = \frac{8}{12}$$

This is the same way dimensional analysis works.

To change the units of a numerical quantity, multiply it by a conversion factor (a fraction with a value equal to 1).

---

example 1

12 feet is equal to how many yards?

The answer to this problem will still have a value equal to 12 <u>feet</u>, but the *label* is being changed to <u>yards</u>.

The hardest part of dimensional analysis is to determine the conversion factor.

In this example, what is the relationship between feet and yards?

3 feet = 1 yard

So now convert this relationship to a fraction equal to 1:

$$\frac{3 \text{ feet}}{1 \text{ yard}} = 1 \qquad \text{and} \qquad \frac{1 \text{ yard}}{3 \text{ feet}} = 1$$

Neither $\frac{3}{1}$ nor $\frac{1}{3}$ may look like a fraction equal to 1, but it's the *units* that make each fraction have a value of 1.

But which one is the conversion factor needed in this problem?

Since the answer is to be in "yards", use the conversion factor with "yards" in the numerator:

$$\frac{12 \text{ feet}}{1} \times \frac{1 \text{ yard}}{3 \text{ feet}}$$

That way, the label "feet" cancels:

$$\frac{12}{1} \times \frac{1 \text{ yard}}{3} = \frac{12}{3} \text{ yards}$$

$$= 4 \text{ yards}$$

So 12 feet = 4 yards.

---

example 2

The Madhany family traveled 3560 miles on a trip across the United States. Since one mile is about 1.6 kilometers, which of the following is closest to the total number of kilometers in 3560 miles?

A. 5700 kilometers
B. 4540 kilometers
C. 3558 kilometers
D. 2225 kilometers

In this problem, miles must be converted to kilometers and the conversion factor needed is given to you right in the problem.

The dimensional analysis can be set up as the original quantity multiplied by a conversion factor:

$$\frac{3560 \text{ miles}}{1} \times \frac{1.6 \text{ kilometers}}{1 \text{ mile}}$$

Next, cancel like units from each numerator and denominator:

$$\frac{3560}{1} \times \frac{1.6 \text{ kilometers}}{1}$$

$$= \frac{3560 \times 1.6 \text{ kilometers}}{1}$$

Multiply 3560×1.6

$$= 5696 \text{ kilometers}$$

So 3560 miles equals approximately 5700 kilometers.

The answer is A.

---

example 3

The largest natural lake in Massachusetts is Assawompsett Pond which has an area of 2,656 acres. What is the approximate area in square miles?

A. 4 sq. mi.
B. 7 sq. mi.
C. 17 sq. mi.
D. 40 sq. mi.

In this problem, acres must be converted to square miles, but the conversion factor needed is not given in the problem. It would need to be provided on a reference sheet or from some other source in order to answer this question. (In this case, you need to know that 1 square mile equals 640 acres.)

The dimensional analysis can now be set up the same way as the previous example:

$$\frac{2656 \text{ acres}}{1} \times \frac{1 \text{ square mile}}{640 \text{ acres}}$$

$$= \frac{2656 \times 1 \text{ square mile}}{1 \times 640}$$

Divide $2656 \div 640$

$$= 4.15 \text{ square miles}$$

So 2,656 acres is approximately 4 square miles.

The answer is A.

---

example 4

If a car is traveling 30 miles per hour, how many feet per minute is the car moving?

This problem changes a *rate*, which has <u>two</u> units of measurement, distance *and* time.

Dimensional analysis can be used to convert units of measure within a rate because a rate can be expressed as a fraction. We'll just need two conversion factors instead of one.

The conversion factors for changing miles to feet and hours to minutes are:

$$\frac{5280 \text{ feet}}{1 \text{ mile}} \quad \text{and} \quad \frac{1 \text{ hour}}{60 \text{ minutes}}$$

Writing this all out so we can cancel like units:

$$\frac{30 \text{ miles}}{1 \text{ hour}} \times \frac{5280 \text{ feet}}{1 \text{ mile}} \times \frac{1 \text{ hour}}{60 \text{ minutes}}$$

$$= \frac{30 \cancel{\text{ miles}}}{1 \cancel{\text{ hour}}} \times \frac{5280 \text{ feet}}{1 \cancel{\text{ mile}}} \times \frac{1 \cancel{\text{ hour}}}{60 \text{ minutes}}$$

$$= \frac{30 \times 5280 \text{ feet} \times 1}{1 \times 1 \times 60 \text{ minutes}}$$

$$= \frac{158400 \text{ feet}}{60 \text{ minutes}}$$

$$= \frac{2640 \text{ feet}}{1 \text{ minute}}$$

A car traveling 30 miles per hour is also moving 2640 feet per minute.

Practice Problems

Name _____

Use dimensional analysis to solve.

1) During lunch, a local fast-food restaurant sold 48 quarter-pound hamburgers. How many kilograms of hamburgers were sold?

2) Bill and Sheila's living room is 18 feet long and 12 feet wide. How many square yards is the area of their living room?

3) A mining company removes 30 tons of rock from a quarry every day. If the company works around the clock, (24 hours per day), how many pounds of rock are removed from the quarry each minute?

4) 1 cubic meter is how many cubic millimeters? Express your answer in scientific notation.

Use dimensional analysis to solve.

5) Deanna's little sister wants to go on a carnival ride, but only children taller than 120 centimeters may go on it. Deanna's sister is 4 feet tall. Can she go on the ride?

6) Light travels about 300,000 kilometers per second. The Sun is approximately 150,000,000 kilometers away from the Earth. How many minutes does it take light from the Sun to reach Earth?

7) Boston and Springfield are 90 miles apart in Massachusetts. How many kilometers apart are the two cities?

8) The distance around the track at the high school is 440 yards. How many meters long is the track?

Lesson 42: Common Geometric Figures

A closed figure made up of line segments is called a *polygon*. Most two-dimensional geometric figures are polygons, and they are named according to the number of sides they have:

Name	Number of Sides	Number of Interior Angles
Triangle	3	3
Quadrilateral	4	4
Pentagon	5	5
Hexagon	6	6
Heptagon	7	7
Octagon	8	8
Nonagon	9	9
Decagon	10	10

One common figure that is missing from the table above is a **circle**, because a circle is *not* a polygon. A circle is a collection of points in a plane (in two-dimensional space) that are all the same distance away from a fixed point. This fixed point is the *center* of the circle, and a line segment connecting the center to any other point on the circle is called a *radius* of the circle:

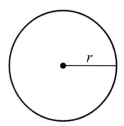

A circle is not a polygon because it is not made up of line segments. Another common geometric figure that is *not* a polygon is an **ellipse**:

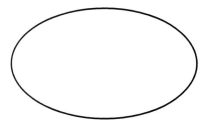

Besides circles and ellipses, just about every other geometric figure you will ever work with will be a polygon. One special type of polygon (first defined in Lesson 36) is a *regular polygon* (which is a polygon whose sides are all congruent and whose interior angles are all congruent).

Triangles

A **triangle** is a three-sided polygon. The sum of the measures of the interior angles of a triangle is 180°.

Types of triangles include:

An **equilateral triangle** (or **equiangular triangle**) is a triangle with all three sides having the same length. The interior angles of an equilateral triangle are also congruent and each measure 60°. An equilateral triangle is also a *regular* triangle.

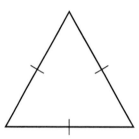

An **isosceles triangle** is a triangle with two sides that are of equal length.

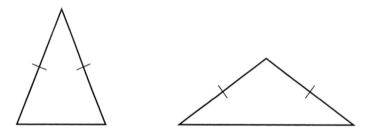

A **scalene triangle** is a triangle with three sides that each have a different length.

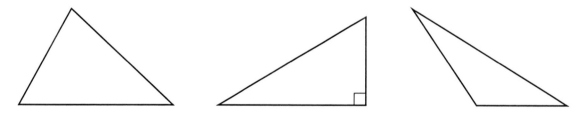

An **acute triangle** is a triangle whose angles are all acute angles. (An acute angle has a measure less than 90°.)

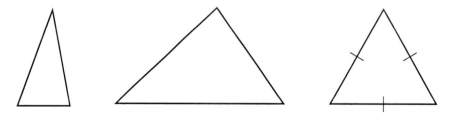

An **obtuse triangle** is a triangle with an obtuse angle. (An obtuse angle has a measure greater than 90º.)

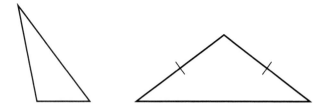

A **right triangle** is a triangle with a right angle. (A right angle has a measure of exactly 90º.)

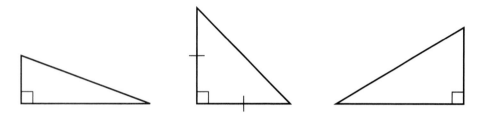

Quadrilaterals

A **quadrilateral** is a four-sided polygon. The sum of the measures of the interior angles of a quadrilateral is 360º.

Types of quadrilaterals include:

A **parallelogram** is a quadrilateral with two pairs of parallel sides.

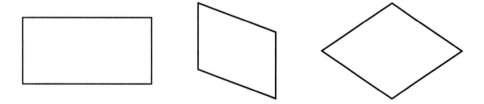

A **rhombus** is a quadrilateral with four sides of equal length.

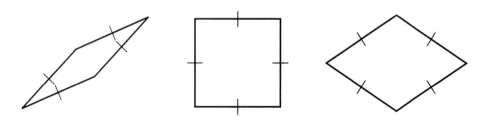

A **rectangle** is a quadrilateral with four interior angles of equal measure, or in other words, all four angles are right angles.

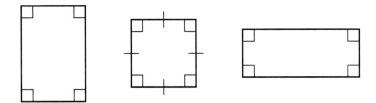

A **square** is a *regular* quadrilateral, or in other words, it is a quadrilateral with four sides of equal length and four interior angles of equal measure (which is 90°).

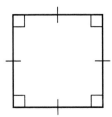

A **trapezoid** is a quadrilateral with exactly one pair of parallel sides. The two parallel sides are called the bases of the trapezoid.

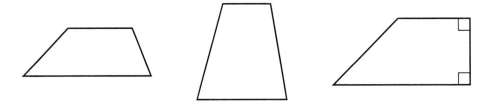

example 1

A partial drawing of a quadrilateral is shown below.

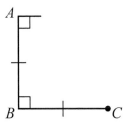

If no other sides or angles are congruent, which **best** describes the figure?

A. square

B. rectangle

C. parallelogram

D. trapezoid

Investigating each answer:

Answer A: To be a square, the other sides *and* angles must be congruent, so the answer cannot be A.

Answer B: To be a rectangle, opposite sides must be congruent, and all angles must be congruent (right angles), so the answer cannot be B.

Answer C: To be a parallelogram, opposite sides must be parallel, but if two adjacent sides are given as being congruent and perpendicular to each other, the only possible parallelogram would be a square, which has already been ruled out, so the answer cannot be C.

Answer D: Answers A, B, and C have already been eliminated, so the answer *must* be D, even if you don't know what a trapezoid is. However, a trapezoid is a quadrilateral (four-sided polygon) with exactly one pair of parallel sides. The given figure could easily be drawn as a trapezoid and meet the criteria listed in the problem (no other sides or angles being congruent)

The answer is D.

---

example 2

Ricardo drew a rectangle. Which statement **must** be true?

A. Ricardo's figure is a parallelogram.
B. Ricardo's figure is a regular polygon.
C. Ricardo's figure is a square.
D. Ricardo's figure is a rhombus.

To answer this question, the definition of *rectangle* must be examined.

A rectangle is a polygon with four right interior angles. This implies that it is also a parallelogram. However, this does *not* mean that all four sides must be congruent (although opposite sides *are* congruent).

Now let's look at the four statements given:

Answer A) A rectangle is, in fact, a parallelogram. But, to be sure this is the answer, the other answers must be checked.

Answer B) A rectangle is not a regular polygon, *unless it is a square*. Remember that all of the sides and interior angles of a regular polygon are congruent, which is true for a square, but isn't always true for all rectangles.

Answer C) A rectangle *can* be a square, but does not *have* to be a square.

Answer D) A rectangle is a rhombus, only when the rhombus is a square. This is because a rhombus is a quadrilateral with four congruent sides, but the interior angles of a rhombus do not *have* to be right angles, although they *can* be.

The answer is A.

Pentagons

A **pentagon** is a five-sided polygon. The sum of the measures of the interior angles of a pentagon is 540º.

regular pentagon

irregular pentagon

Hexagons

A **hexagon** is a six-sided polygon. The sum of the measures of the interior angles of a hexagon is 720º.

regular hexagon

irregular hexagon

Octagons

An **octagon** is an eight-sided polygon. The sum of the measures of the interior angles of a octagon is 1080º.

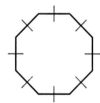
regular octagon

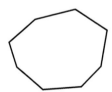
irregular octagon

N-gons

An ***n*-gon** is an *n*-sided polygon. The sum of the measures of the interior angles of an *n*-gon is $(n - 2) \times 180º$.

Practice Problems

Name _____

Match the appropriate label below with each quadrilateral. Some figures will require more than one label.

A) parallelogram
B) rhombus
C) rectangle
D) square
E) trapezoid

1)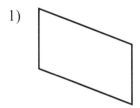

2)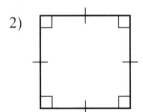

3)

4)

Match the appropriate label below with each triangle. Some figures will require more than one label.

A) equilateral triangle
B) isosceles triangle
C) scalene triangle
D) acute triangle
E) obtuse triangle
F) right triangle

5)
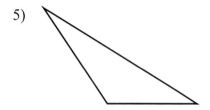

6)
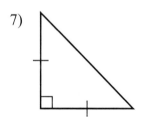

7)
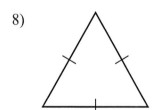

8)
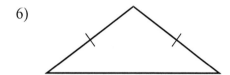

9)
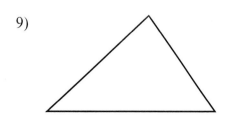

Lesson 43: Perimeter of Common Geometric Figures

The **perimeter** of a figure is the sum of the lengths of all its sides.

<u>Formulas for the perimeter of common figures</u>:

square: $P = 4s$

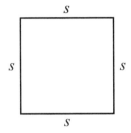

s = length of each side of a square

rectangle: $P = 2w + 2l$

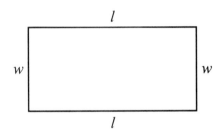

l = length of the rectangle
w = width of the rectangle

circle: The "perimeter" of a circle is called the **circumference** of the circle.

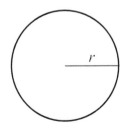

The formula for circumference is: $C = 2\pi r$, where r is the radius of the circle and $\pi \approx 3.14$. (Remember, π is an irrational number with a decimal value that goes on indefinitely; 3.14 is just an approximation.)

example 1

Find the perimeter of the given polygon.

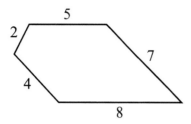

$P = 2 + 4 + 8 + 7 + 5$

$P = 26$ units

example 2

Find the circumference of a circle with a radius of 1 cm.

$C = 2\pi r$

$C = 2\pi(1 \text{ cm})$

$C = (2 \text{ cm})\pi$

$C = (2 \text{ cm})(3.14)$

$C = 6.28 \text{ cm}$

example 3

Which of the following **best** describes the meaning of π?

A. the measurement of the circumference of a circle
B. the measurement of the diameter of a circle
C. the ratio of the circumference of a circle to its diameter
D. the ratio of the area of a circle to its circumference

Pi (π) is a symbol you see when dealing with circles, either when finding the circumference or area of a circle (and also with spheres, cones, or any other rounded shape). Usually you are told just to know that $\pi \approx 3.14$, but where does that number originate?

The formula for the circumference of a circle is $C = 2\pi r$, where r is the radius of the circle. If the radius is 1, the circumference is $2 \times \pi \times 1$, or 2π. Why is there a 2 in this formula?

When it comes to measuring a circle, the radius is the distance from the center of the circle to the circle itself, and the diameter is the distance from one side of the circle to the other side (as long as you pass through the center):

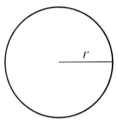 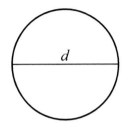

Because the diameter is twice the length of the radius, that is where the 2 comes from in the formula for circumference:

$$C = 2\pi r \qquad \text{or} \qquad C = \pi d$$

If this formula is solved for π:

$$C = \pi d$$

$$\frac{C}{d} = \frac{\pi d}{d}$$

$$\frac{C}{d} = \pi$$

From this, we can see that π is equal to the ratio of the circumference of a circle to the diameter of that circle.

The answer is C.

---

example 4

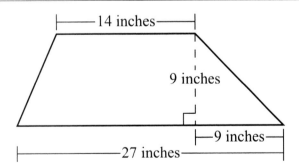

Which measure is closest to the perimeter of the trapezoid?

A. 41 inches

B. 59 inches

C. 64 inches

D. 66 inches

Before finding the perimeter, the length of each side of the trapezoid must be found. The trapezoid can be broken up into a rectangle and two right triangles.

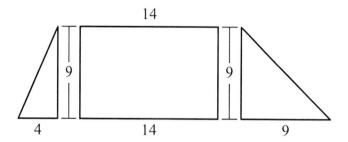

The entire base of the trapezoid is 27 inches, so if the triangle to the right has a base of 9 inches and the rectangle has a length of 14 inches, then the triangle to the left must have a base of 4 inches ($27 = 9 + 14 + 4$).

Using the Pythagorean Theorem, the length of each hypotenuse of the two right triangles can be found, and these are the two measures needed to find the overall perimeter of the trapezoid.

left triangle	right triangle
$a^2 + b^2 = c^2$	$a^2 + b^2 = c^2$
$4^2 + 9^2 = c^2$	$9^2 + 9^2 = c^2$
$16 + 81 = c^2$	$81 + 81 = c^2$
$97 = c^2$	$162 = c^2$
$c \approx 9.85$	$c \approx 12.73$

$$P = 14 + 9.85 + 27 + 12.73$$

$$P = 63.58$$

The answer is C.

Practice Problems

Name _____

Find the perimeter of each figure.

1)

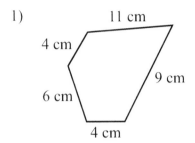

2)

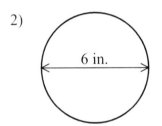

3)

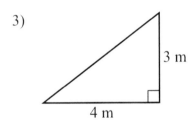

4)

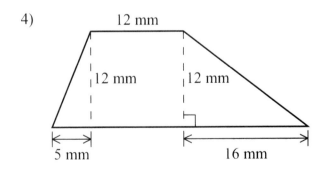

5) The radius of the Earth is approximately 4,000 miles. How long is the Equator?

6) Complete the following table.

Rectangles

length	width	Perimeter
10 meters		30 meters
14 feet	12 feet	
	8 inches	20 inches

Find the perimeter of each figure.

7)

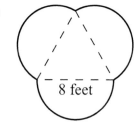

8 feet

8)

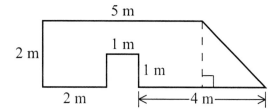

9)

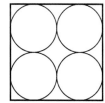

Each circle has a radius of 3 inches.

Lesson 44: Area of Common Geometric Figures

The **area** of a figure is the measure of the size of the region enclosed by the figure.

Formulas for the area of common figures:

square: $A = s^2$

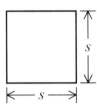

s = length of each side of a square

rectangle: $A = l \cdot w$

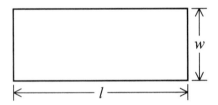

l = length of a rectangle
w = width of a rectangle

parallelogram: $A = b \cdot h$

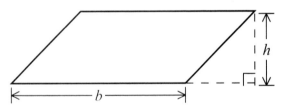

b = base measure of the parallelogram
h = height of the parallelogram

triangle: $A = \dfrac{1}{2} b \cdot h$

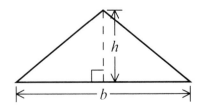

b = base measure of a triangle
h = height of a triangle

trapezoid: $A = \dfrac{1}{2}h(b_1 + b_2)$

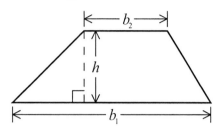

h = height of a trapezoid
b_1 = measure of the top of the trapezoid (base 1)
b_2 = measure of the bottom of the trapezoid (base 2)

circle: $A = \pi r^2$

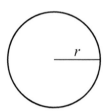

r = radius of a circle

---

example 1

What is the area of the **shaded** region in the figure below? (Use 3.14 for π.)

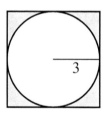

A. 5.15 sq. units

B. 17.16 sq. units

C. 7.74 sq. units

D. 21.72 sq. units

In the above figure, a circle is inscribed in a square (which means the circle touches every side of the square). Because the diameter of the circle is the same length as each side of the square, the dimensions of the square can be found by doubling the radius of the circle. (Remember that the radius of a circle is half of the diameter.) So if the circle has a radius of 3 units, the length of each side of the square is 6 units.

To find the area of the shaded region, the area of the square must be found, as well as the area of the circle. The area of the circle can then be subtracted from the area of the square:

Area of the square   Area of the circle

$A = s^2$ $A = \pi r^2$

$A = (6 \text{ units})^2$ $A = (3.14)(3 \text{ units})^2$

$A = 36 \text{ square units}$ $A = (3.14)(9 \text{ square units})$

                                             $A = 28.26 \text{ square units}$

Now subtract the area of the circle from the area of the square:

Area of square – Area of circle = Shaded Area

36 sq. units – 28.26 sq. units = 7.74 sq. units

The answer is C.

---

example 2

What is h, the height of the parallelogram represented below, if its area is 91 square centimeters.

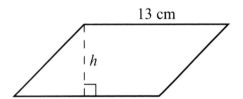

A. 7 cm
B. 9 cm
C. 11 cm
D. 15 cm

The formula for the area of a parallelogram is $A = b \cdot h$ (base × height).

$A = b \cdot h$

$$\frac{91 \text{ cm}^2}{13 \text{ cm}} = \frac{(13 \text{ cm}) \cdot h}{13 \text{ cm}}$$

$h = 7 \text{ cm}$

The answer is A.

example 3

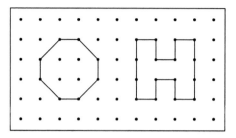

Which statement is true about the two polygons?

A. They have equal areas but different perimeters.
B. They have equal perimeters but different areas.
C. They have equal areas and equal perimeters.
D. They have different areas and different perimeters.

Since the answer to this question depends on how the areas and perimeters of the two polygons compare, both will need to be calculated for each polygon:

The polygon on the left, shaped like an octagon, has 8 sides. Four of the sides have a length of 1 unit, but the sides that are diagonal are a little longer than that. You can imagine each of them as being the hypotenuse of a right triangle with legs that have a length of 1.

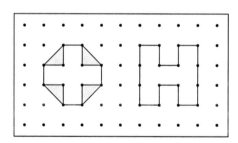

This means each "hypotenuse" would have a length of $\sqrt{2}$, or about 1.4:

$$a^2 + b^2 = c^2 \quad \Rightarrow \quad 1^2 + 1^2 = c^2 \quad \Rightarrow \quad c^2 = 2 \quad \Rightarrow \quad c = \sqrt{2} \text{ or } {\sim}1.4$$

So with four sides with a length of 1, and four sides with a length of about 1.4, the perimeter of the polygon on the left is around 9.6 units.

The polygon on the right is made up of all straight line segments, each with a length of 1 unit. You can count these line segments, almost like a game of "connect the dots", and count 16 of them. So, the perimeter of the polygon on the right is 16 units. This means the two polygons have different perimeters.

To find the area of each polygon, draw additional line segments inside each one, dividing them into squares that are 1 unit by 1 unit.

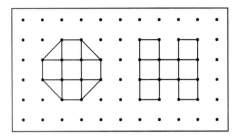

The figure on the left would end up having 5 of these squares, and 4 right triangles, each of which is half of the area of a square. So, the area of the left polygon is 7 square units (5 squares + 4 half-squares, or 5 square units + 2 square units).

The polygon on the right ends up having 7 of these squares, so the area of this polygon is also 7 square units.

It turns out that the polygons have different perimeters but equal areas.

The answer is A.

---

example 4

Two rectangles, *ABCD* and *WXYZ*, are shown below. The measure of each side of *WXYZ* is 5 times the measure of each corresponding side of *ABCD*.

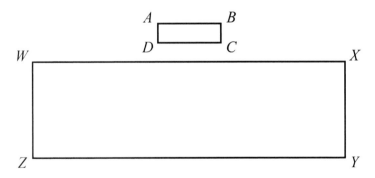

Which statement is true of the areas of these two rectangles?

A. The area of *WXYZ* is 5 times the area of *ABCD*.
B. The area of *WXYZ* is 10 times the area of *ABCD*.
C. The area of *WXYZ* is 20 times the area of *ABCD*.
D. The area of *WXYZ* is 25 times the area of *ABCD*.

The area of a rectangle can be found be multiplying the length of the rectangle by its width:

Area of rectangle = length × width, or, $A = l \cdot w$

Since the length of the larger rectangle is 5 times the measure of the smaller rectangle, and the width is also 5 times as big, we can replace the *l* and *w* in the area formula with a *5l* and a *5w*:

Area of larger rectangle = 5 times the length × 5 times the width

or, $A = 5l \cdot 5w$

Now we have a formula for comparing the area of the larger rectangle to the area of the smaller rectangle. The associative property of multiplication (see Lesson 1) allows the formula to be re-written with the numbers and variables grouped differently:

$A = (5 \cdot 5)(l \cdot w)$

which can be simplified to:

$A = 25(l \cdot w)$

So the area of the larger rectangle would be 25 times the area of the smaller rectangle.

The answer is D.

Practice Problems

Name _____

Find the area of each figure.

1)

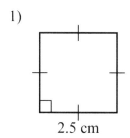

2.5 cm

2)

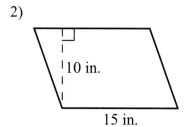

10 in.

15 in.

3)

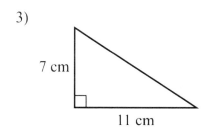

7 cm

11 cm

4)

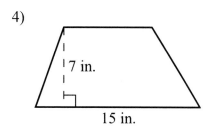

7 in.

15 in.

5)

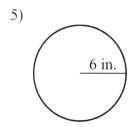

6 in.

6)

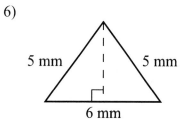

5 mm   5 mm

6 mm

7) The perimeter of the figure below is 24 units. What is the area of the figure?

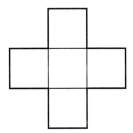

8) The area of the figure below is 8 square units. What is the perimeter of the figure?

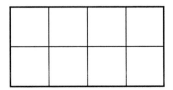

Find the area and perimeter of the figures given below.

9)

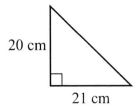

20 cm

21 cm

10)   10 in.

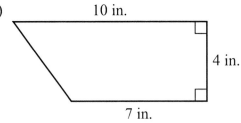

4 in.

7 in.

11) Complete the following table.

Rectangles

length	width	Perimeter	Area
2	6		
9			45
	8	36	
	4		28

Lesson 45: Three-Dimensional Figures

A three-dimensional figure is made up of points that do not all lie in the same plane (a **plane** is a flat surface that extends forever two-dimensionally in all directions.)

Some of these are called **polyhedrons**, which are closed three-dimensional figures formed by flat surfaces that are polygons. Some three-dimensional figures are not polyhedrons, such as spheres, cones, or cylinders, because they have curved surfaces.

Spheres

A **sphere** is a three-dimensional figure made up of points that are all a set distance from a fixed point in space.

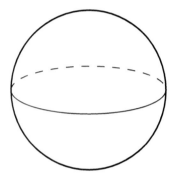

Cones

A **cone** is a three-dimensional figure with one vertex and a circular base. (A **base** is a bottom side of a geometric figure.)

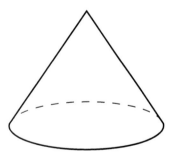

Cylinders

A **cylinder** is a three-dimensional figure with two parallel bases that are congruent circles.

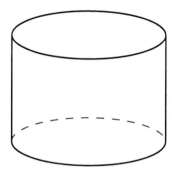

Prisms

A **prism** is a three-dimensional figure with two parallel bases that are congruent polygons and whose faces are rectangles.

Prisms are named for the shape its bases. Types of prisms include:

A **triangular prism** is a prism with congruent triangular bases.

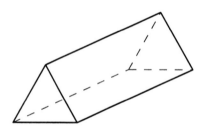

A **rectangular prism** is a prism with congruent rectangular bases.

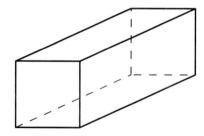

If the bases *and* faces of a rectangular prism are squares, then that prism is a **cube**.

(Note: The prisms above are shown with their bases at the front and back of the diagrams instead of at the top and bottom.)

Pyramids

A **pyramid** is a three-dimensional figure with a polygon for its base and whose faces are triangles having a common vertex.

Pyramids are named for the shape of its base. Types of pyramids include:

A **triangular pyramid** is a pyramid with a triangular base.

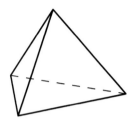

A **square pyramid** is a pyramid with a square base.

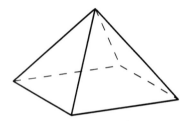

<table>
<tr><td>example 1</td></tr>
</table>

What is the **minimum** number of congruent, equilateral triangles needed to construct a three-dimensional figure if no other shapes are used?

A. 3
B. 4
C. 6
D. 8

Just as a polygon is a closed figure and made up of line segments, a polyhedron is a region (or volume) enclosed by polygons. [*Polyhedron* comes from the Greek words *polus* (meaning "many") and *hedra* (meaning "base", "seat", or "face").]

Every polygon that makes up a polyhedron shares each of its edges with another polygon, and also shares each of its vertices with two other polygons. So, since each triangle has three sides (or edges) that must be shared with another triangle, that would a total of four triangles needed to make a polyhedron (which would be a triangular pyramid in this case.

The answer is B.

---

example 2

Mei Ling gave the following description of a three-dimensional figure.

- The solid has 6 faces.

- The solid has 8 vertices.

- The solid has 12 edges.

Which of the following figures matches Mei Ling's description?

A.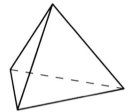

C.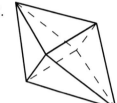

B.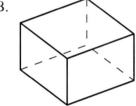

D.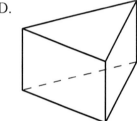

To better understand this question, you need to know the definition of *face*, *vertex* (*vertices* is the plural of *vertex*) and *edge*:

- a **face** is a flat surface of a three-dimensional figure, or in other words, one of the polygons that makes up the figure

- an **edge** is a line segment where two faces of a three-dimensional figure meet

- a **vertex** is a point where three or more edges of a three-dimensional figure meet

So a face is a surface, an edge is a line segment, and a vertex is a point. Now we just need to see which of the figures given has 6 surfaces, 12 line segments, and 8 points.

Answer A shows a **tetrahedron** (*tetra*- means four), which has 4 faces, 6 edges, and 4 vertices. Answer B shows a rectangular prism, which has 6 faces, 12 edges, and 8 vertices. Answer C shows an **octahedron** (*octa*- means eight), which has 8 faces, 12 edges, and 6 vertices. Answer D shows a triangular prism, which has 5 faces, 9 edges, and 6 vertices.

The answer is B.

Geometric Nets

A **net** is the two-dimensional representation of a three-dimensional figure that has been unfolded.

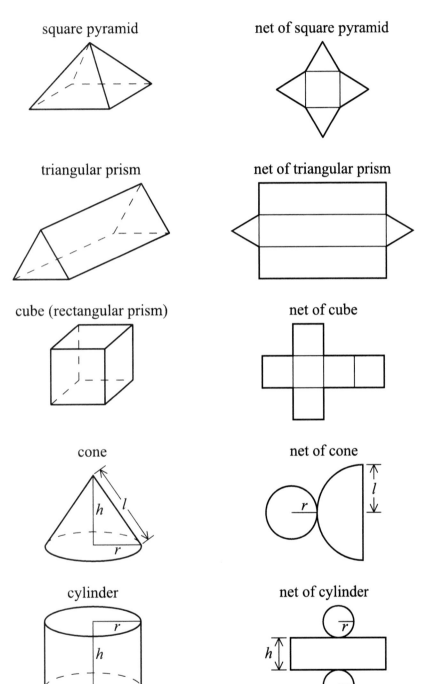

square pyramid

net of square pyramid

triangular prism

net of triangular prism

cube (rectangular prism)

net of cube

cone

net of cone

cylinder

net of cylinder

example 3

Which of the following patterns could be folded to form a square pyramid?

A.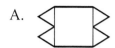

B.

C.

D.

A pyramid is a polyhedron, or closed three-dimensional figure made up of polygons, and a square pyramid is a pyramid with a square base.

Answer A shows four triangles surrounding a square, but two sides of the square share edges with *two* triangles, and two sides of the square share edges with *zero* triangles. This can't make a polyhedron.

Answer B is eliminated as a possible answer because none of the polygons that make it are squares, and the base must be a square.

Answer C is also eliminated because it can't make a polyhedron, which is made of polygons that share edges and vertices. Answer C is made of polygons with one vertex of each triangle connected to each edge of the square, but not connected to each vertex of the square.

Answer D has a square, with each side sharing an edge with a triangle. If these triangles were folded upward, they would be able to share edges with each other and have a common vertex at the top of the pyramid.

The answer is D.

Practice Problems

Name _____

Name each three-dimensional figure shown and list the number of vertices, edges, and faces that figure has.

1)

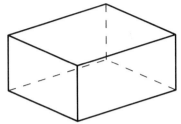

2)

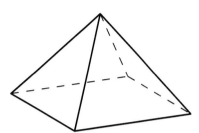

3)

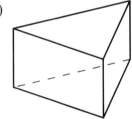

4)

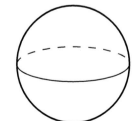

5)

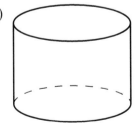

6)

Draw the geometric net for each three-dimensional figure.

7)

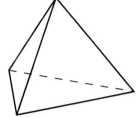

8)

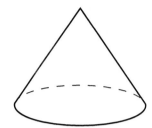

Draw the three-dimensional figure that each given geometric net represents. If a three-dimensional figure cannot be drawn from the given net, explain why.

9)

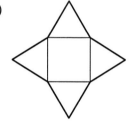

10)

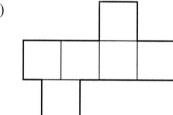

11)

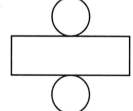

12)

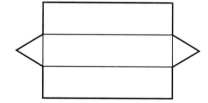

Lesson 46: Surface Area of Three-Dimensional Figures

The **surface area** of a three-dimensional figure is the sum of the areas of all of its surfaces. The **lateral area** of a three-dimensional figure is the surface area excluding the area of any bases (the top or bottom of a three-dimensional figure).

Note: Area is measured in square units.

<u>Spheres</u>

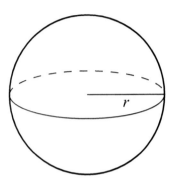

Surface Area (S.A.) = $4\pi r^2$ r = radius of a sphere

example 1

Find the surface area of the given sphere.

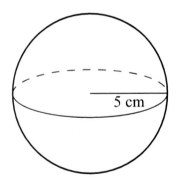

S.A. = $4\pi r^2$

S.A. = $4\pi(5 \text{ cm})^2$

S.A. = $4\pi(25 \text{ cm}^2)$

S.A. = $100\pi \text{ cm}^2$

S.A. = $100(3.14) \text{ cm}^2$

S.A. = 314 cm^2

<u>Cones</u>

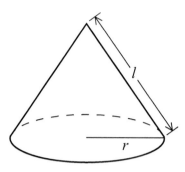

Lateral Area (L.A.) = $\pi r l$

S.A. = $\pi r l + \pi r^2$

r = radius of the base of a cone

l = slant height of a cone

πr^2 is the area of the base of the cone.

$\pi r l$ is the area of the lateral face of the cone.

---

example 2

Find the lateral area and surface area of the given cone.

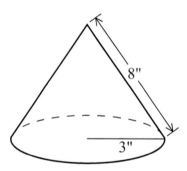

L.A. = $\pi r l$

L.A. = π(3 in.)(8 in.)

L.A. = π(24 in.^{2})

L.A. = (3.14)(24 in.^{2})

L.A. = 75.36 in.^{2}

S.A. = $\pi r l + \pi r^2$

S.A. = π(3 in.)(8 in.) + π(2 in.)^2

S.A. = π(24 in.^{2}) + π(4 in.^{2})

S.A. = 28π in.^{2}

S.A. = 28(3.14) in.^{2}

S.A. = 87.92 in.^{2}

Cylinders

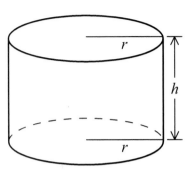

L.A. = $2\pi rh$ r = radius of a cylinder
S.A. = $2\pi r^2 + 2\pi rh$ h = height of a cylinder

πr^2 is the area of the top or bottom of the cylinder, so $2\pi r^2$ is the area of the top and bottom added together.

$2\pi rh$ is the area of the lateral face of the cylinder, which would look like a label pealed off a soup can and flattened.

example 3

Find the lateral area and surface area of the given cylinder.

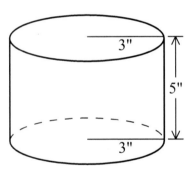

L.A. = $2\pi rh$ S.A. = $2\pi r^2 + 2\pi rh$

L.A. = $2\pi(3 \text{ in.})(5 \text{ in.})$ S.A. = $2\pi(3 \text{ in.})^2 + 2\pi(3 \text{ in.})(5 \text{ in.})$

L.A. = $2\pi(15 \text{ in.}^2)$ S.A. = $2\pi(9 \text{ in.}^2) + 2\pi(15 \text{ in.}^2)$

L.A. = $30\pi \text{ in.}^2$ S.A. = $18\pi \text{ in.}^2 + 30\pi \text{ in.}^2$

L.A. = $30(3.14) \text{ in.}^2$ S.A. = $48\pi \text{ in.}^2$

L.A. = 94.20 in.^2 S.A. = $48(3.14) \text{ in.}^2$

                                             S.A. = 150.72 in.^2

Prisms

The type of prism depends on the polygon that makes up the bases of the prism. (The lateral faces of any prism are rectangles.)

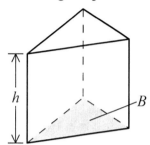

triangular prism

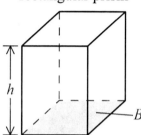

rectangular prism

L.A. = $P \cdot h$
S.A. = $P \cdot h + 2B$

P = base perimeter of a prism
h = height of a prism
B = base area of a prism

---

example 4

Find the lateral area and surface area of the given prism.

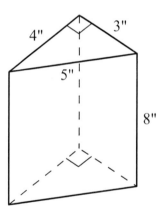

L.A. = $P \cdot h$

P = 3 in. + 4 in. + 5 in.

P = 12 in.

L.A. = (12 in.)(8 in.)

L.A. = 96 in.^{2}

S.A. = $P \cdot h + 2B$

$B = \dfrac{1}{2} b \cdot h$

$B = \dfrac{1}{2} (4 \text{ in.})(3 \text{ in.}) = 6 \text{ in.}^2$

S.A. = (12 in.)(8 in.) + 2(6 in.^{2})

S.A. = 96 in.^{2} + 12 in.^{2}

S.A. = 108 in.^{2}

---

example 5

Find the lateral area and surface area of the given prism.

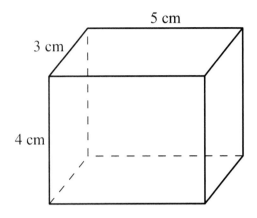

L.A. $= P \cdot h$

$\quad P = 3$ cm $+ 5$ cm $+ 3$ cm $+ 5$cm

$\quad P = 16$ cm

L.A. $= (16$ cm$)(4$ cm$)$

L.A. $= 64$ cm^2

S.A. $= P \cdot h + 2B$

$\quad B = l \cdot w$

$\quad B = (3$ cm$)(5$ cm$) = 15$ cm^2

S.A. $= (16$ cm$)(4$ cm$) + 2(15$ cm$^2)$

S.A. $= 64$ cm$^2 + 30$ cm^2

S.A. $= 94$ cm^2

Pyramids (Regular Pyramids)

A **regular pyramid** has a base that is a regular polygon and has lateral faces that are isosceles triangles.

examples of regular polygons

regular quadrilateral
(also called a *square*)

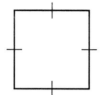

regular triangle
(also called an *equilateral triangle*)

regular pentagon

regular hexagon

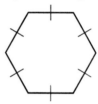

$$\text{L.A.} = \frac{1}{2}P \cdot l$$ P = base perimeter of a pyramid

$$\text{S.A.} = \frac{1}{2}P \cdot l + B$$ l = slant height

B = base area of a pyramid

Practice Problems

Name _____

Find the lateral area of each of the given three-dimensional figures.

1)

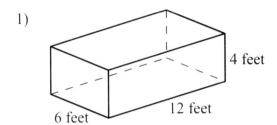

4 feet

12 feet

6 feet

2)

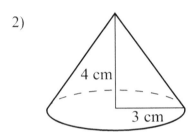

4 cm

3 cm

3)

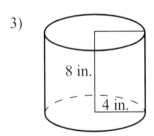

8 in.

4 in.

4)

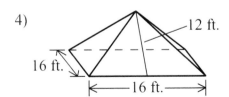

12 ft.

16 ft.

16 ft.

Find the surface area of each of the given three-dimensional figures.

5)

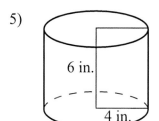

6)

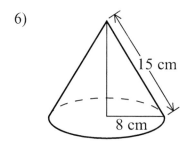

7)

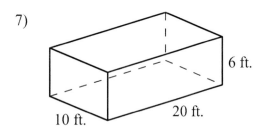

8)

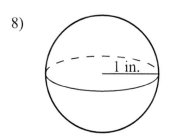

9)

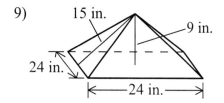

Lesson 47: Volume of Three-Dimensional Figures

The **volume** of a three-dimensional figure is the measure of the amount of space it encloses.

Note: Volume is measured in cubic units.

Spheres

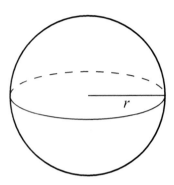

$$\text{Volume} = \frac{4}{3}\pi r^3 \qquad\qquad r = \text{radius of a sphere}$$

example 1

Find the volume for the given sphere.

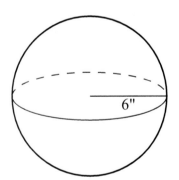

$$V = \frac{4}{3}\pi r^3$$

$$V = \frac{4}{3}\pi (6 \text{ in.})^3$$

$$V = \frac{4}{3}\pi (216 \text{ in.}^3)$$

$$V = 288\pi \text{ in.}^3$$

$$V = 288(3.14) \text{ in.}^3$$

$$V = 904.32 \text{ in.}^3$$

Cones

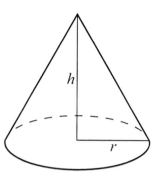

Volume $= \dfrac{1}{3}\pi r^2 h$

$r =$ radius of a cone

$h =$ height of a cone

example 2

Find the volume of the given cone.

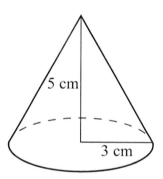

$V = \dfrac{1}{3}\pi r^2 h$

$V = \dfrac{1}{3}\pi (3 \text{ cm})^2 (5 \text{ cm})$

$V = \dfrac{1}{3}\pi (9 \text{ cm}^2)(5 \text{ cm})$

$V = \dfrac{1}{3}\pi (45 \text{ cm}^3)$

$V = \dfrac{1}{3}(3.14)(45 \text{ cm}^3)$

$V = 47.10 \text{ cm}^3$

Cylinders

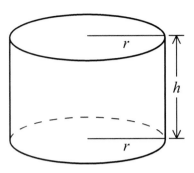

Volume = $\pi r^2 h$ r = radius of a cylinder
                                               h = height of a cylinder

example 3

Find the volume of the given cylinder.

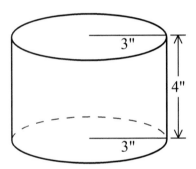

$V = \pi r^2 h$

$V = \pi(3 \text{ in.})^2(4 \text{ in.})$

$V = \pi(9 \text{ in.}^2)(4 \text{ in.})$

$V = \pi(36 \text{ in.}^3)$

$V = (3.14)(36 \text{ in.}^3)$

$V = 113.04 \text{ in.}^3$

Prisms

In the last section, we saw that there are many kinds of prisms, with each kind defined by the polygon shape of the base.

triangular prism   rectangular prism

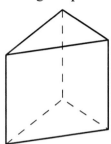   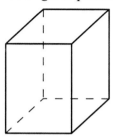

The volume of *any* prism is the area of the base times the height of the prism:

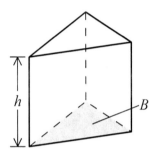   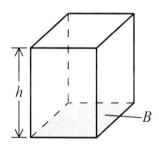

Volume = $B \cdot h$

   B = base area of a prism

   h = height of a prism

For a triangular prism, $B = \dfrac{1}{2} b \cdot h$ (the area of a triangle), and for a rectangular prism, $B = l \cdot w$

Since a triangular prism can have *any* kind of triangle for its base, how you find the area may vary (depending on what information you're given to start with, and from that, how you define b and h when you determine the area of that triangle), so the formula for a triangular prism is always expressed as:

Volume = $B \cdot h$

example 4

Find the volume of the given triangular prism.

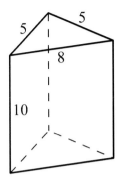

$V = B \cdot h$

First find the area of the base.

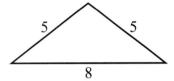

The base is an isosceles triangle that can be divided into two right triangles.

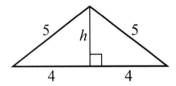

The Pythagorean Theorem can be used to find h.

$$a^2 + b^2 = c^2$$

$$h^2 + 4^2 = 5^2$$

$$h^2 + 16 = 25$$
$$-16 \; -16$$

$$h^2 = 9$$

$$h = 3 \text{ units}$$

Now we know that $b = 8$ and $h = 3$ for the triangle, so the base area can be found:

$$B = \frac{1}{2}b \cdot h$$

$$B = \frac{1}{2}(8)(3)$$

$$B = 12 \text{ square units}$$

Now substitute this into the formula for volume:

$$V = B \cdot h$$

$$V = (12 \text{ square units})(10 \text{ units})$$

$$V = 120 \text{ cubic units}$$

For the base area of a rectangular prism, B equals $l \times w$, the area of *any* rectangle. This can be substituted for B in the formula for the volume of a rectangular prism:

$$V = B \cdot h \quad \text{becomes} \quad V = l \cdot w \cdot h$$

example 5

Find the volume of the given rectangular prism.

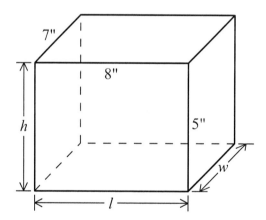

$$V = l \cdot w \cdot h$$

$$V = (8 \text{ in.})(7 \text{ in.})(5 \text{ in.})$$

$$V = (56 \text{ in.}^2)(5 \text{ in.})$$

$$V = 280 \text{ in.}^3$$

Pyramids

As with prisms, there are many types of pyramids, depending on the polygon shape of the base (which is typically a regular polygon).

square pyramid

triangular pyramid

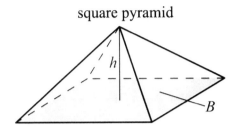
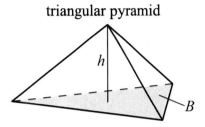

For *all* pyramids:

$$\text{Volume} = \frac{1}{3}B \cdot h$$

B = base area of a pyramid

h = height of a pyramid

---

example 6

Find the volume of the given pyramid.

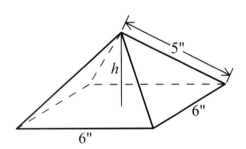

$V = \dfrac{1}{3}B \cdot h$

B = the area of a square

$B = s^2$

$B = (6 \text{ in.})^2$

$B = 36 \text{ in.}^2$

To find *h*, first find the length of a diagonal of the base:

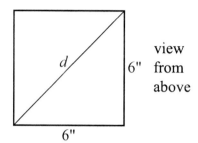

view
6" from
above

6"

$$a^2 + b^2 = c^2$$

$$(6 \text{ in.})^2 + (6 \text{ in.})^2 = d^2$$

$$36 \text{ in.}^2 + 36 \text{ in.}^2 = d^2$$

$$72 \text{ in.}^2 = d^2$$

$$d = \sqrt{72} \text{ or } 8.49 \text{ in.}$$

Now use this measurement to find *h*.

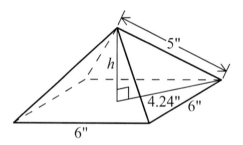

With a slant height of 5, and using half the length of the diagonal, we can visualize a right triangle within the pyramid. Using the Pythagorean Theorem, we can find *h*:

$$a^2 + b^2 = c^2$$

$$h^2 + (4.24 \text{ in.})^2 = (5 \text{ in.})^2$$

$$h^2 + 17.98 \text{ in.}^2 = 25 \text{ in.}^2$$
$$ -17.98 \text{ in.}^2 -17.98 \text{ in.}^2$$

$$h^2 = 7.02 \text{ in.}^2$$

$$h = \sqrt{7.02} \text{ or } 2.65 \text{ in.}$$

Now the volume can be calculated.

$$V = \frac{1}{3} B \cdot h$$

$$V = \frac{1}{3} (36 \text{ in.}^2)(2.65 \text{ in.})$$

$$V = 31.80 \text{ in.}^3$$

Practice Problems

Name _____

Find the volume of each of the given three-dimensional figures.

1)

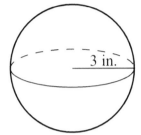

3 in.

2)

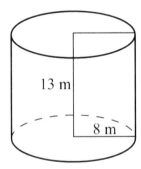

13 m

8 m

3)

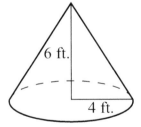

6 ft.

4 ft.

Find the volume of each of the given three-dimensional figures.

4)

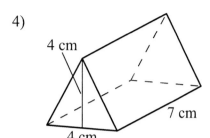

5)

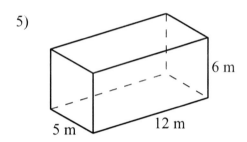

6)

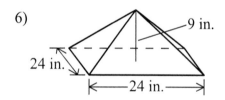

7)

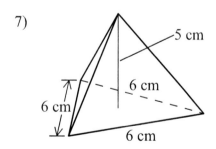

Data Analysis and Probability Standard for Grades 6–8
Expectations

Instructional programs from prekindergarten through grade 12 should enable all students to—	In grades 6–8 all students should—	The following lessons correspond to each expectation—
Formulate questions that can be addressed with data and collect, organize, and display relevant data to answer them	• formulate questions, design studies, and collect data about a characteristic shared by two populations or different characteristics within one population; • select, create, and use appropriate graphical representations of data, including histograms, box plots, and scatterplots.	• Lesson 58 • Lessons 48, 50, 51, 52, 53, 54, and 57
Select and use appropriate statistical methods to analyze data	• find, use, and interpret measures of center and spread, including mean and interquartile range; • discuss and understand the correspondence between data sets and their graphical representations, especially histograms, stem-and-leaf plots, box plots, and scatterplots.	• Lessons 55 and 56 • Lessons 48, 50, 51, 52, 53, 54, and 57
Develop and evaluate inferences and predictions that are based on data	• use observations about differences between two or more samples to make conjectures about the populations from which the samples were taken; • make conjectures about possible relationships between two characteristics of a sample on the basis of scatterplots of the data and approximate lines of fit; • use conjectures to formulate new questions and plan new studies to answer them.	• Lesson 58 • Lessons 48 and 49 • no matching lessons
Understand and apply basic concepts of probability	• understand and use appropriate terminology to describe complementary and mutually exclusive events; • use proportionality and a basic understanding of probability to make and test conjectures about the results of experiments and simulations; • compute probabilities for simple compound events, using such methods as organized lists, tree diagrams, and area models.	• Lesson 61 • Lesson 59 • Lesson 61

Lesson 48: Scatterplots and Stem-and-Leaf Plots

<u>Scatterplots</u>

Any two sets of numbers can be matched together into ordered pairs (as long as each set has the same number of terms). These two sets may be completely independent from each other, or they may have some kind of relationship.

---

example 1

Karen's mother is a veterinarian, and the table below shows the number of cats being held overnight at the animal hospital on each of Karen's birthdays over five years.

Karen's age	number of cats
6	3
7	8
8	5
9	2
10	7

The number of cats Karen's mom would need to keep overnight at the animal hospital has *nothing* to do with Karen's age, but we can make ordered pairs from these sets of numbers anyway:

$$\{(6, 3), (7, 8), (8, 5), (9, 2), (10, 7)\}$$

If we plot these ordered pairs on a graph:

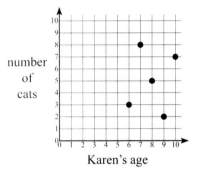

A graph of plotted ordered pairs is called a **scatterplot** because the points are "scattered" around the graph.

The purpose of a scatterplot is to see if any relationship between the two sets of data appears that wasn't obvious from looking at the list of ordered pairs.

The relationship shown on a scatterplot is called a **correlation**.

When the pattern shown by a set of points on a scatterplot has a *positive slope*, the scatterplot shows a **positive correlation**.

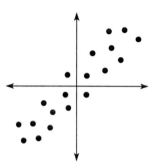

When the pattern shown by a set of points on a scatterplot has a *negative slope*, the scatterplot shows a **negative correlation**.

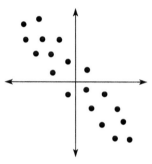

If the points on a scatterplot show no pattern at all (as in the previous example of Karen's age vs. number of cats), the graph displays **no correlation**.

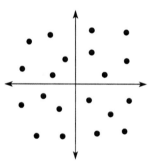

example 2

Steve went for a 6-mile run and checked his watch after every mile. The results are listed in the table below:

number of miles run	number of minutes elapsed
1	6.0
2	12.5
3	19.5
4	26.0
5	33.0
6	40.5

Plotting these ordered pairs gives the following scatterplot:

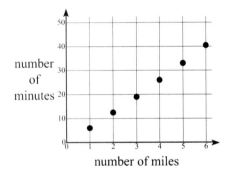

In this scatterplot, an obvious positive correlation can be seen. The scatterplot shows that the longer Steve ran, the more distance he covered.

example 3

In the scatter plot, each dot represents one student who participated in the 50-meter race. Ben is 15 years old. Based on the information on the scatter plot, what was Ben's time in the race?

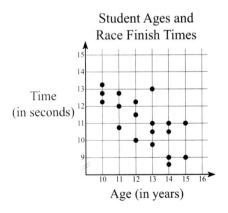

A. 9 seconds

B. 10 seconds

C. 11 seconds

D. It cannot be determined.

Looking at the scatterplot, you can see that there were three 10-year-olds, three 11-year-olds, three 12-year-olds, four 13-year-olds, four 14-year-olds, and two 15-year-olds that were in the race. One of the 15-year-olds had a time of 11 seconds, and the other had a time of 9 seconds. Which one of the 15-year-olds was Ben? The scatterplot doesn't display that information, so it's impossible to know.

The answer is D.

Stem-and-Leaf Plots

There are many ways to display a list of numbers, and a **stem-and-leaf plot** is one of them. The reasons for using a stem-and-leaf plot are:

▶ It organizes your data in increasing or decreasing order.

▶ The lowest and greatest values and the range of the data (the difference between the greatest and lowest values) are easy to identify.

▶ Any values that repeat, or any patterns in how the numbers are grouped, can also be easily identified.

The following example will go through the steps for making a stem-and-leaf plot:

example 4

Make a stem-and-leaf plot for the given set of numbers.

{42, 17, 39, 25, 15, 33, 40, 19, 36, 22}

1) Find the greatest common place value of all the numbers in the data set.

The greatest common place value for this set of numbers is the *tens place*.

2) Write out the *stems* of the stem-and-leaf plot. The stems are the digits from the common place value of the data set.

stem	leaf
1	
2	
3	
4	

3) Write out the *leaves* of the stem-and-leaf plot, which come from the digits in the next place value down from the stems. In other words, in this example, since the digits from the tens place became the stems, the digits from the ones place will be the leaves. (Note: The leaves can be written in either ascending or descending order.)

stem	leaf
1	5 7 9
2	2 5
3	3 6 9
4	0 2

4) Write the **key** next to the stem-and-leaf plot. The key lets others know what place value the stems and the leaves came from.

stem	leaf
1	5 7 9
2	2 5
3	3 6 9
4	0 2

$1 \mid 5 = 15$ ← key

Now that the set of data is in a stem-and-leaf plot, it is easy to see that the lowest value is 15, the greatest value is 42, the range is 27 (42 minus 15), no values repeat, and there are no obvious grouping of values within the data.

If two additional values are added to the original set of numbers, we get the following:

$$\{42, 17, 39, 25, 15, 33, 40, 19, 36, 22, 6, 107\}$$

Now the greatest common place value is the *ones place* because of the number 6, but the same stem-and-leaf plot can be created by using a stem of "0" for that data value, and a stem of "10" can be used for the number 107.

stem	leaf
0	6
1	5 7 9
2	2 5
3	3 6 9
4	0 2
10	7

$1 \mid 5 = 15$

Note: Stem values can be skipped if there are no numbers in the data set with corresponding digits in that place value. In the above example, stems 5 through 9 were skipped because there were no numbers in the data set in the 50s, 60s, 70s, 80s, or 90s.

Because only two digits from each number in a data set can be displayed in a stem-and-leaf plot, numbers with more than two digits must be rounded off.

---

example 5

Display the following data in a stem-and-leaf plot:

$$\{143, 261, 179, 133, 204, 307, 157, 162, 312\}$$

The greatest common place value is the hundreds place, so the leaves will come from the tens place. This means that each number must be rounded to the nearest ten. (See Lesson 40 for the rules on rounding off numbers.)

After rounding, we get a new data set:

$$\{140, 260, 180, 130, 200, 310, 160, 160, 310\}$$

These can now be written in a stem-and-leaf plot:

stem	leaf
1	3 4 6 6
2	0 6
3	1 1

$1 \mid 3 = 130$

Remember, without the key, anyone looking at this stem-and-leaf plot wouldn't know that the stems come from the hundreds place and the leaves come from the tens place.

Because numbers displayed in a stem-and-leaf plot can only show the first two digits of any numbers, if you have any data that cannot be rounded, some other method of displaying data should be chosen.

Practice Problems

Name _____

For the scatterplots below, label whether each has a *positive*, *negative*, or *no correlation*.

1) _____

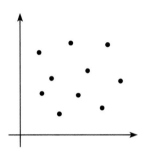

2) _____

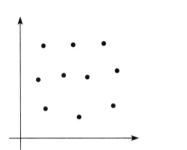

3) _____

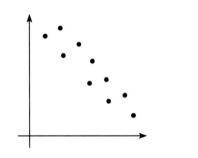

4) _____

5) Make a scatterplot by plotting the following set of ordered pairs:

$$\{(4, 3), (7, 10), (5, 6), (7, 7), (1, 4), (3, 6), (6, 8), (2, 2), (9, 8), (6, 4)\}$$

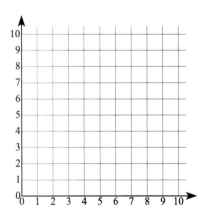

6) Make a stem-and-leaf plot from the data given in the table below.

24	35	26	18	28	47
13	42	56	33	25	27
40	34	26	42	11	32

7) Make a stem-and-leaf plot from the data given in the table below.

680	770	880	820	830	700
790	930	640	770	830	750
690	880	780	620	790	920
940	970	750	880	750	810

Use the stem-and-leaf plot below to answer questions 8–10.

Daily High Temperatures for October

stem	leaf
3	8 9
4	3 6 7 8 9
5	3 4 6 7 7 8 8 9
6	0 1 2 2 2 4 5 5 6 6 7 8
7	0 1 4

$3 \mid 8 = 38$

8) What was the lowest high temperature that occurred in October?

9) On how many days was the high temperature in the 50s?

10) What high temperature occurred the most often?

Lesson 49: Approximating a Line of Best Fit

When ordered pairs are created from a linear equation, those points can be plotted with a line then drawn through them to represent the linear equation on a graph.

When a scatterplot is drawn, the points are usually *not* linear, but if there is a correlation (positive or negative) visible in the pattern of the points, a line can be sketched through the scatterplot to represent the *trend* of the scatterplot. This **trend line** is called a **line of best fit**.

The line on the graph below best represents all of the points in the scatterplot:

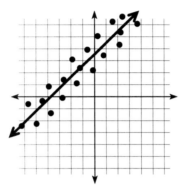

Note: A line of best fit is an approximation, so it's possible that two different people will draw two different lines on the same scatterplot.

When drawing a line of best fit:

► Try to have just as many points above the line as there are below the line.

► Try to draw the line through as many of the "middle" points as possible.

Once a line of best fit is drawn, finding the *equation* of that line is important, as it allows predictions to be made about information beyond the scatterplot.

---

example 1

Mrs. Kramer asked her students to report the number of hours they studied for their statistics test. The day after the test, she plotted the results on the scatterplot shown below.

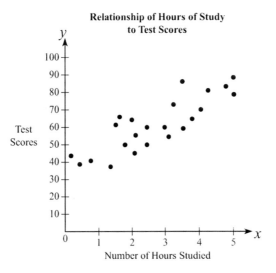

Which of the following equations correctly approximates the line of best fit?

A. $y = -10x + 30$

B. $y = -10x + 60$

C. $y = 10x + 30$

D. $y = 10x + 60$

Notice how the answers are written in slope-intercept form ($y = mx + b$), where m is the slope of the line, and b is the y-intercept.

A line of best fit might look like this:

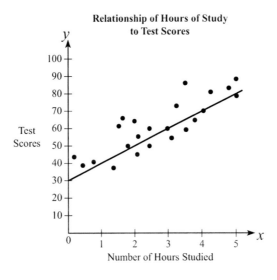

The slope of this line ("rise" over "run") has an approximate vertical increase of 10 for each horizontal increase of 1, so $m = 10$. The line also looks like it would cross the y-axis at approximately 30.

Putting these numbers in the slope-intercept equation for m and b gives:

$$y = mx + b$$

$$y = 10x + 30$$

The answer is C.

Another way to determine the equation of a line is to take two points from the scatterplot that your line passes through or near.

From the above example, $(3, 60)$ and $(5, 80)$ would be two such points.

The point-slope equation of a line can be used for this:

$$y - y_1 = m(x - x_1)$$

x_1 and y_1 are the coordinates of any point on a line, and m is the slope of that line.

First calculate the slope:

$$m = \frac{y_2 - y_1}{x_2 - x_1}$$

$$m = \frac{80 - 60}{5 - 3}$$

$$m = \frac{20}{2}$$

$$m = 10$$

Then use either point, $(3, 60)$ or $(5, 80)$, as x_1 and y_1:

let $x_1 = 3$ and $y_1 = 60$

$$y - y_1 = m(x - x_1)$$

$$y - 60 = 10(x - 3)$$

$$\begin{array}{l} y - 60 = 10x - 30 \\ +60 +60 \end{array}$$

$$y = 10x + 30$$

THIS PAGE INTENTIONALLY BLANK

Practice Problems

Name _____

For questions 1–3, determine the equation of the line of best fit for each scatterplot.

1)

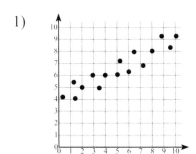

2)

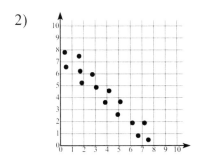

3)

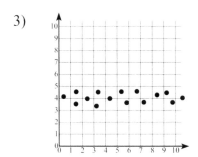

For questions 4 and 5, plot the given ordered pairs, draw a line of best fit on your scatterplot, and then determine the equation of the line of best fit that you sketched.

4)

x	1	2	3	2	2	1	3	2	3	3	1
y	3	2	17	18	7	11	20	10	6	12	0

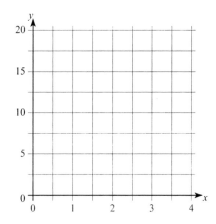

5)

x	4	6	2	1	8	5	6	3	8	0	1
y	2	0	4	2	1	1	3	3	0	4	3

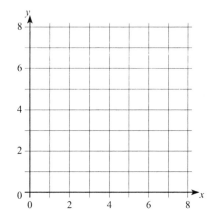

Lesson 50: Circle Graphs

A table or stem-and-leaf plot are ways of listing numbers in an organized manner, but if you want to make a visual representation of data, *graphs* can be used.

If you want to make a visual comparison between data values, a **circle graph** is good to use.

A circle graph is a circle that represents the total number of values (or items, etc.) in the data set. The circle is then divided into slices that are proportional to the ratio of the number of one item to the total number of items.

The table below shows what type of sneaker each student was wearing in a gym class.

type of sneaker	number of students with that type
cross-training	3
running	6
basketball	5
tennis	2
skateboarding	4

To create a circle graph with this data:

1) First find the total number of sneakers:

$3 + 6 + 5 + 2 + 4 = 20$

2) Determine what fraction of the circle will be represented by each type of sneaker:

cross-training: $\dfrac{3}{20}$

running: $\dfrac{6}{20}$

basketball: $\dfrac{5}{20}$

tennis: $\dfrac{2}{20}$

skateboarding: $\dfrac{4}{20}$

Note: When these fractions are added together, the total should be 1.

$$\frac{3}{20} + \frac{6}{20} + \frac{5}{20} + \frac{2}{20} + \frac{4}{20} = \frac{20}{20} = 1$$

3 out of 20 students were wearing cross-trainers, 6 out of 20 were wearing running shoes, and so on. So $\frac{3}{20}$ -ths of the circle will be a slice representing cross-training sneakers, but the slices will be easier to draw if you determine how big the *angle* of each slice will be.

3) Convert the previously found fractions to decimals, and then to angle measurements.

To convert to decimals, use a calculator (or long division if you have to):

cross-training: $\frac{3}{20} = 0.15$

running: $\frac{6}{20} = 0.30$

basketball: $\frac{5}{20} = 0.25$

tennis: $\frac{2}{20} = 0.10$

skateboarding: $\frac{4}{20} = 0.20$

Note: When these decimals are added together, the total should be 1.

$$0.15 + 0.30 + 0.25 + 0.10 + 0.20 = 1$$

To convert these decimals to angle measurement, remember that there are 360° in a circle:

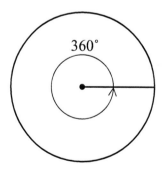

360°

Multiplying each decimal by 360° will give you the angle size of each slice of the circle:

cross-training: $0.15 \times 360° = 54°$

running: $0.30 \times 360° = 108°$

basketball: $0.25 \times 360° = 90°$

tennis: $0.10 \times 360° = 36°$

skateboarding: $0.20 \times 360° = 72°$

4) Draw a circle with slices that have the angle measures you just found:

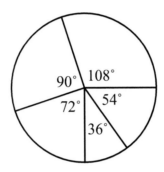

The slices are often shaded or colored, and a key can be shown nearby to label each slice, or the labels can be written directly on the circle graph.

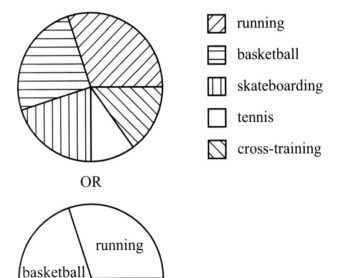

In this circle graph, "running" has the biggest slice and "tennis" has the smallest slice.

Circle graphs are good for making comparisons among the values in a list of data because you can compare the size of each slice to the others.

Note: A circle graph can represent any total number of items, but exactly how many items there are or how many of each item each slice represents cannot be seen on a circle graph unless specifically labeled that way:

---

example 1

Mr. Hong took a survey in which he asked each of the 30 students in his math class to choose his or her favorite food. He tallied the results in this chart.

Favorite Food

Food	Number of Students Choosing Food
Tacos	||||	
Pizza	|||| ||||
French Fries	|||
Hot Dogs	||||
Hamburgers	|||| |||

Mr. Hong wants to construct a circle graph for this data. What should be the measure of the angle for the section of the graph for hamburgers?

A. 40°

B. 56°

C. 120°

D. 96°

A circle graph can represent any number of items. The size of the slices is based on how many items each slice represents in relation to the total number of items. In this case, the items to be represented are students, and there are a total of 30 of them. So the entire circle will represent 30 items (students), and each slice of the circle will represent each set of students based on what favorite food they chose.

Now that we know how many students the entire graph will represent, how many students will each slice represent? According to the table, 6 students chose tacos, 9 students chose pizza, 3 students chose french fries, 4 students chose hot dogs, and 8 students chose hamburgers. The problem is asking about hamburgers, so we need to calculate how the number of students that chose hamburgers compares to the total number of students, which is 8 out of 30.

8 out of 30 is 26.7% of the students (8 ÷ 30 = 0.2666…). But we don't need to know the percentage for how many students chose hamburgers, we need the size of the angle of the slice of the circle graph that would represent these students.

Since there are 360° in a circle, the angle for the section of the graph for hamburgers will be 26.7% of 360°.

$$26.7\% \times 360° = (0.267)(360°) \approx 96°$$

You could also make a fraction from 8 out of 30 and multiply that by 360°.

$$\frac{8}{30} \times \frac{360°}{1} = \frac{2880°}{30} = 96°$$

The answer is D.

---

example 2

The circle graph below shows the percentages of people who brought each type of baked good to sell at a recent bake sale.

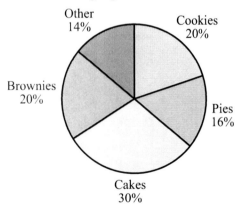

**Percentage of People
Bringing Baked Goods**

Other
14%

Cookies
20%

Brownies
20%

Pies
16%

Cakes
30%

If 15 people brought cakes to sell, what is the total number of people who brought baked goods to sell at the bake sale?

A. 45

B. 50

C. 70

D. 75

To answer this problem, you first need to make sure you understand what each number provided in the problem represents. On the circle graph, instead of labeling the measure of each angle in degrees (with the entire circle having 360º), each slice is labeled by a percentage out of 100. And the number you are given in the problem is not the total number of items represented by the entire circle, but what one slice of the graph represents.

According to the graph, 30% of the people brought cakes. We are also told that the *number* of people who brought cakes is 15. So, 15 is 30% of the total number of people who brought baked goods. This can be written as an algebraic expression:

$$15 \text{ is } 30\% \text{ of the total}$$

$$15 = 30\% \text{ of } x$$

$$15 = 0.30 \cdot x$$

The total number of people is the answer to this question, so x will be the answer we are looking for:

$$15 = 0.30 \cdot x$$

$$\frac{15}{0.30} = \frac{0.30 \cdot x}{0.30}$$

$$50 = x$$

The answer is B.

Practice Problems

Name _____

1) How many degrees is the measure of the angle of each slice of the circle graph below?

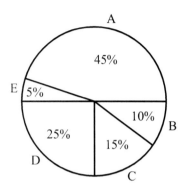

A _____   B _____   C _____   D _____   E _____

2) What percentage of the total number of items represented by the circle graph below does each slice represent?

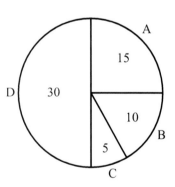

A _____   B _____   C _____   D _____

Use the given values to sketch a circle graph that represents those number. Be sure to calculate the percentage of the total that each item represents and the size of the angle of each slice within the circle graph.

3) Item A: 25
    Item B: 30
    Item C: 15
    Item D: 30

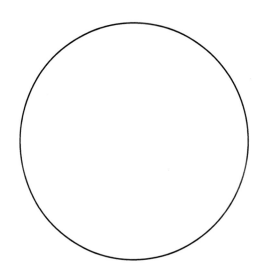

4) Item A: 60
    Item B: 15
    Item C: 100
    Item D: 30
    Item E: 45

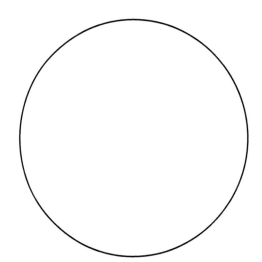

Lesson 51: Line Graphs

While a circle graph is a good way to display a comparison of data, a **line graph** is a good way to display a *change in a value (or values) over time.*

---

example 1

The population of the United States each decade since 1900 is given in the table below:

Year	1900	1910	1920	1930	1940	1950	1960	1970	1980	1900	2000
Pop*	76.2	92.2	106.0	123.2	132.2	151.3	179.3	203.3	226.5	248.7	281.4

*Population given in millions

This same information can be displayed as a line graph:

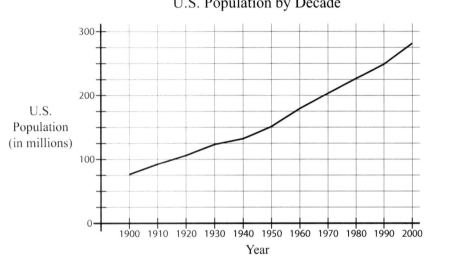

U.S. Population by Decade

A line graph is sometimes called a **broken-line graph** because it is made up of line segments and not one continuous line.

Line graphs are useful for displaying rates. A rate is a ratio of two different quantities, as the example above compares population with time.

The **rate of change** of the population with time can be calculated for each decade, over several decades, or from the first population figure on the graph to the last population plotted.

example 2

What was the rate of population change from 1930 to 1940?

$$\text{rate of change} = \frac{\text{population in 1940} - \text{population in 1930}}{1940 - 1930}$$

$$= \frac{132.2 \text{ million} - 123.2 \text{ million}}{10 \text{ years}}$$

$$= \frac{9 \text{ million}}{10 \text{ years}}$$

$$= 0.9 \text{ million per year}$$

From 1930 to 1940, the average rate of change of the U.S. population was an increase of 900,000 people per year.

example 3

What was the rate of population change from 1990 to 2000?

$$\text{rate of change} = \frac{\text{population in 2000} - \text{population in 1990}}{2000 - 1990}$$

$$= \frac{281.4 \text{ million} - 248.7 \text{ million}}{10 \text{ years}}$$

$$= \frac{32.7 \text{ million}}{10 \text{ years}}$$

$$= 3.27 \text{ million per year}$$

From 1990 to 2000, the average rate of change of the U.S. population was an increase of 3,270,000 people per year.

As shown in the two above examples, the graph shows how the U.S. population grew slowly during the 1930s and grew rapidly during the 1990s.

example 4

The chart below shows the average monthly price per share of HiTek stock for each month in 2001.

2001 HiTek Stock Prices

Month	Jan	Feb	Mar	Apr	May	June	July	Aug	Sep	Oct	Nov	Dec
Average Price (in $)	22.61	24.25	31.02	27.31	29.92	33.10	36.14	35.50	34.01	31.05	36.20	40.12

Which of the following graphs **best** models the general behavior of the stock's price for 2001?

A.

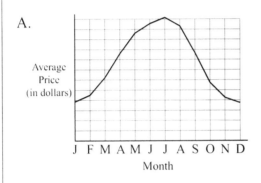

C.

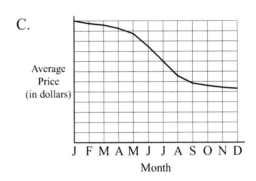

B.

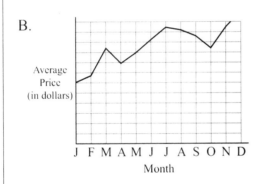

D.

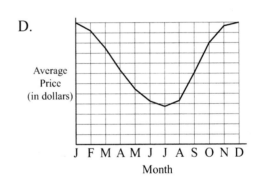

Following the numbers in the table, we can see that the price increased from January to March, decreased from March to April, increased from April to July, decreased from July to October, and increased from October to December. So we're looking for a graph that displays a pattern of up, then down, then up, then down, and finishing by going up.

Without even looking at the actual dollar values, we can see that only one graph displays this pattern of increasing and decreasing amounts over time, which is graph B.

To be sure, let's look at the patterns shown on the other three graphs. Graph A shows an increase for half the year and a big decrease for the second half of the year. Graph C shows one continuous decrease. Finally, Graph D shows a decrease for the first seven months and then an increase to December.

The answer is B.

example 5

On January 25, the treasurer of the Sports Club opened a savings account for the club. The following graph shows the amount of money in the club's account during February and March.

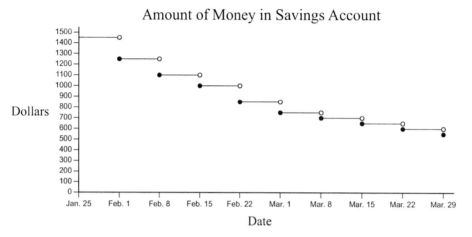

Amount of Money in Savings Account

Based on the information in the graph, which of the following statements is true?

A. More than half of the money was taken out of the account during February.
B. More than half of the money was taken out of the account during March.
C. More money was taken out of the account each week during February than each week during March.
D. More money was taken out of the account each week during March than each week during February.

The above graph displays something called a **step function**, or **discontinuous function**, and is called that because 1) it is a function (notice how it passes the vertical line test), and 2) its value remains constant as x increases, until it jumps suddenly to another value, which then remains constant until the next jump. (This information won't help us solve the problem, but now you know what a step function is.)

Since each statement refers to how the amount of money in the account changed in February and March, we should look at the graph to see how much money was in the account at the beginning and end of those months.

The account was opened with $1450. On February 1st, there was approximately $1250 in the account. On March 1st, the amount was around $750. And on March 29th, the savings account had about $550 in it. This means that $500 was taken out of the account in February and $200 was withdrawn in March.

Both $500 and $200 are less than half of the amount of money that was in the account at the start of each month, so Answers A and B are eliminated. Now it's just a matter of during which month more money was taken out.

The answer is C.

Practice Problems

Name _____

1) The number of cars that passed through a toll booth at the end of each hour on a particular day is shown in the table below. Sketch this information as a line graph.

time	6am	7am	8am	9am	10am	11am	12pm	1pm	2pm	3pm	4pm	5pm	6pm
# of cars	25	48	63	52	34	30	32	38	31	35	46	58	67

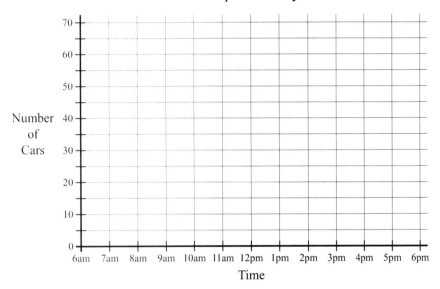

Number of Cars per Hour by a Toll Booth

a) Using the line graph you made above, what 2-hour period had the biggest increase in the number of cars that passed through the toll booth?

b) What 2-hour period had the biggest decrease in the number of cars?

c) Based on the line graph, would the number of cars to pass through the toll booth in the next hour (6pm to 7pm) be larger or smaller than the number of cars between 5pm and 6pm (67)?

d) Based on common sense, do you think that would really happen?

Distance Traveled Over Time

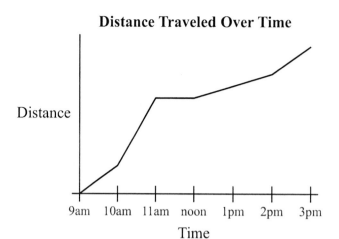

2) The above line graph represents a driving trip taken by Gene.

a) During what 1-hour period was Gene likely driving on a highway?

b) During what hours did he stop driving?

c) During what hours was Gene driving, but at his slowest rate of the trip?

U.S. National Debt

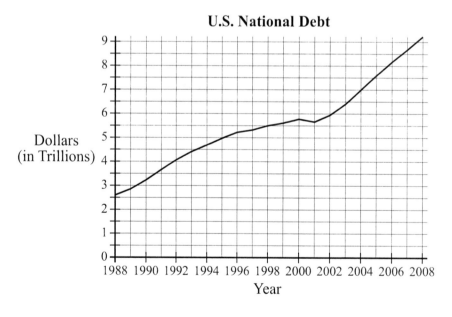

3) Based on the line graph above, during what period of time has the National Debt increased the fastest?

Lesson 52: Line Plots

Another method for displaying numerical information is with a line plot.

A **line plot** is basically a number line with data points plotted *above* it.

---

example 1

Twenty students in an algebra class got the following scores on a test:
78, 87, 91, 72, 68, 93, 75, 76, 84, 90, 78, 83, 85, 72, 95, 70, 78, 83, 89, 91

To display this data on a line plot:

Step 1) Determine the **greatest value** and **lowest value** and draw a number line that begins just before the lowest value and ends just after the greatest value. (The end points are often multiples of 5, 10, or some other increment).

       lowest value: 68
       greatest value: 95

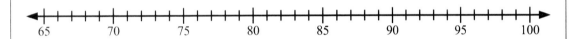

Step 2) Draw an "×" or point "•" *above* the number line to represent each value in the list of numbers. If values repeat, draw the second "×" directly above the first one, and so on.

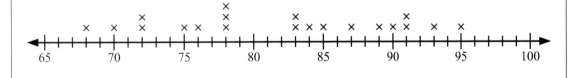

A line plot gives a visual representation of data that makes it easy to:

▶ see the greatest and lowest values

▶ see what values repeat and which one(s) repeat the most

▶ see any grouping patterns or gaps in the data

In the line plot above, it's easy to see that the lowest value is 68, the greatest value is 95, the range of the scores is 27 (95 – 68 = 27), the test score of 78 occurred the most often, and the data fell into two groups (one group between 68 and 78, and another group between 83 and 95).

example 2

The line plot below shows the number of people in each student's household for a class of students.

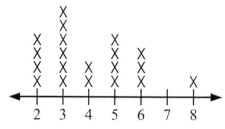

Number of People in
Student Households

What is the mean number of people in households for this class of students?

A. 3

B. 3.5

C. 4

D. 6

The *mean* of a set of numbers is the *average* of those numbers. (This will be explained more in Lesson 55.) To find the average, list all of the values displayed in the line plot, add them up, and then divide by the number of values that were added:

2, 2, 2, 2, 3, 3, 3, 3, 3, 3, 4, 4, 5, 5, 5, 5, 6, 6, 6, 8   →   20 values (1 from each of the 20 students in the class)

$$\frac{2+2+2+2+3+3+3+3+3+3+4+4+5+5+5+5+6+6+6+8}{20} = \frac{80}{20}$$

$$= 4$$

The mean number of people in each student household is 4.

The answer is C.

Practice Problems

Name _____

1) Twenty math grades are listed below:

    76, 85, 93, 73, 79, 83, 87, 70, 76, 85, 81, 76, 89, 70, 81, 78, 85, 68, 76, 89

  a) Make a line plot of the grades.

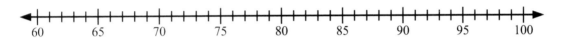

  b) What is the range of the list of grades?

  c) What grade appears on the line plot more than any of the others?

  d) What is the lowest grade?

  e) What is the highest grade?

  f) What 10-point range contains most of the grades (such as from 60-70, 85-95, etc.)?

2) A list of values is represented on the line plot below.

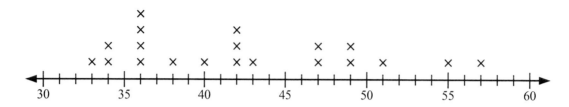

a) Make a list of all of the numbers represented on the above line plot.

b) What is the lowest value on the line plot?

c) What is the greatest value?

d) What is the range of these values?

e) What value repeats the most?

Lesson 53: Bar Graphs

While line graphs are useful for showing a change in data over time or rates of change among two different quantities, **bar graphs** are useful for displaying and comparing data (especially when "time" is not one of the quantities, although it still can be).

example 1

One hundred people were surveyed about their favorite color. The results are shown in the table below.

color	number of votes
blue	29
red	22
green	14
orange	9
black	20
purple	6

This information can be presented as a bar graph:

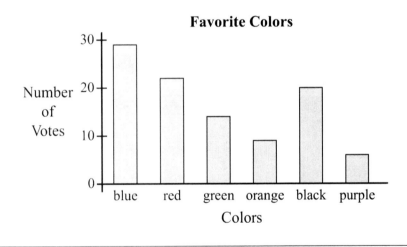

A few rules to follow when making a bar graph:

▶ make all bars the same width

▶ use an equal amount of space between each bar

▶ label each *bar*, and label each *axis*

▶ always try to begin each numerical axis at zero

▶ give the bar graph a title

351

Data on a bar graph can be displayed in any order. If order *is* important, a line graph might be a better choice, unless a *comparison* between the data is the most important aspect of the graph (in which case a circle graph may also work).

---

example 2

A comparison of the median wage for a business employee in three cities is shown below.

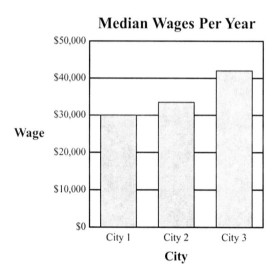

Which of the following is closest to the difference between the median wage in City 2 and the median wage in City 1?

A. $4,000

B. $8,000

C. $12,000

D. $64,000

The key to answering this problem is being able to interpret the value represented by each bar on the graph. The bar above "City 1" represents $30,000, which is easy to see because the top of the bar is right at the line for $30,000. The value for the other two bars, however, needs to be estimated. And actually, we only need to estimate the value for City 2 in order to answer this problem. That value is between $30,000 and $40,000, and looks to be closer to $30,000. So a guess of $33,000 or $34,000 would be all we need.

The next thing to do is to find the *difference* between the two values:

$$\$33,000 - \$30,000 = \$3,000 \quad \text{or} \quad \$34,000 - \$30,000 = \$4,000$$

So the difference is around $3,000 or $4,000.

The answer is A.

Practice Problems

Name _____

1) Carl Yastrzemski played for the Boston Red Sox from 1961 to 1983. The number of homeruns he hit over a five-year period is shown in the table below. Sketch a bar graph from this information. (Remember to label everything on your graph.)

Year	1967	1968	1969	1970	1971
Homeruns	44	23	40	40	15

a) What was the total number of homeruns hit during that 5-year period?

b) What is the percent of change in the number of homeruns hit from 1968 to 1969?

c) How many more homeruns were hit in 1970 than in 1971?

d) What percentage of all homeruns hit during this period occurred in 1967?

2) A group of students were asked to name their favorite weekday. Plot the following results in a bar graph. (Be sure to label everything on your graph.)

Day	Monday	Tuesday	Wednesday	Thursday	Friday
# of Votes	2	3	5	7	13

3) The bar graph below represents the grades that students received on an exam.

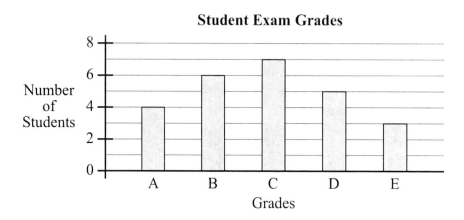

a) How many students took this exam?

b) What percentage of students failed the exam?

c) What might be a better way to display the student scores if you wanted to see exactly what each student grade was?

Lesson 54: Histograms and Venn Diagrams

Frequency Distribution Charts

In a set of data, sometimes there are values that repeat. The number of times a data value occurs in a data set is the **frequency** of that data value.

A **frequency distribution chart** is a table that lists how often data values repeat in a data set.

example 1

The following data set can be rewritten in a frequency distribution chart:

$$\{4, 9, 6, 5, 4, 6, 4, 8, 5, 9, 6, 4, 7, 6\}$$

data value	frequency
4	4
5	2
6	4
7	1
8	1
9	2

Histograms

Sometimes data is divided into ranges, such as grades, like how 90-100 is an A, 80-89 is a B, 70-79 is a C, etc.

A frequency distribution chart can be made using these intervals.

example 2

Rewrite the following grades in a frequency distribution chart:

84, 77, 91, 75, 78, 85, 93, 67, 71, 94, 66, 81, 62, 96, 87, 70, 92, 68, 75, 76

grade interval	frequency
60-69	4
70-79	7
80-89	4
90-100	5

When a frequency distribution chart involving intervals is graphed as a bar graph, it is called a **histogram**.

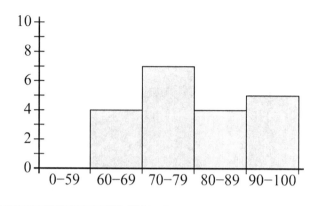

There are a couple of differences between a bar graph and a histogram:

▶ On a histogram, both axes are numerical (where the vertical axis measures the frequency and the horizontal axis is labeled with the intervals).

▶ On a histogram there are *no* spaces between the bars.

Venn Diagrams

A **Venn diagram** is a series of circles (or other geometric figures) that displays the relationship between various sets of data.

example 3

23 students are currently taking Pre-Algebra and 18 students are taking History. 11 of the students are in both classes. Display this information in a Venn diagram.

Just like with a circle graph, one circle can represent an entire set of data.

To begin, draw one circle to represent the Pre-Algebra students and another circle to represent the History students:

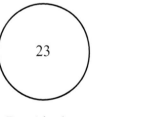
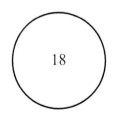

Pre-Algebra   History

The circles can be drawn with one overlapping the other to represent the students in both classes:

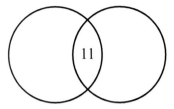

Pre-Algebra   History

To complete this Venn diagram, the number of students taking *only* Pre-Algebra or History needs to be calculated:

The number of students taking *only* Pre-Algebra equals the total number of Pre-Algebra students minus the number of students in both classes:

Pre-Algebra only students = 23 – 11 = 12

The number of students taking *only* History equals the total number of History students minus the number of student in both classes:

History only students = 18 – 11 = 7

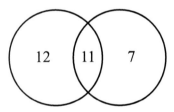

Pre-Algebra   History

Note: It is important that the numbers within each circle add up to the *total* number of items that circle represents.

example 4

The Venn diagram below shows Leila's graduating classes from middle school, high school, and college.

Leila's Graduating Class

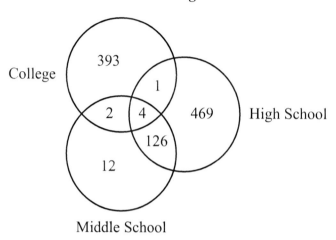

How many students graduated together from **both** Leila's middle school and high school?

A. 133
B. 132
C. 131
D. 130

The area where the circle for Leila's high school overlaps with the circle for Leila's middle school is what represents students that went to both schools. Some of these students went only to her high school and middle school, but not to the same college as her, while other students went to all three schools with her.

The area where *only* the high school circle and middle school circle overlap is what represents students that went only to those two schools, while the area where all three circles overlap is what represents students that graduated from Leila's middle school, high school, and college.

The Venn diagram shows that 126 students graduated from her middle school and her high school (but then didn't go to the same college as Leila), and that 4 students graduated from all three schools. That is a total of 130 students that graduated together from both Leila's middle school and high school.

The answer is D.

Practice Problems

Name _____

1) Fill in the frequency distribution chart based on the following list of numbers:

12, 38, 19, 24, 29, 30, 11, 36, 17, 23, 27, 32, 31, 18, 19, 25, 28, 22, 24, 33

Interval	Frequency
11 – 15	
16 – 20	
21 – 25	
26 – 30	
31 – 35	
36 – 40	

2) Draw a histogram of the above data.

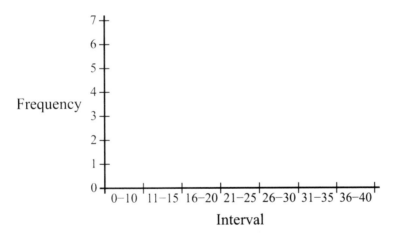

3) What percentage of the numbers given above are less than or equal to 20?

4) What interval contains the greatest amount of numbers?

5) The number of eighth graders that play sports in the fall, winter, and spring is shown in the Venn diagram below.

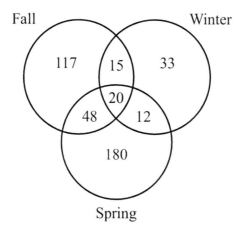

a) How many students play sports in the fall?

b) How many students play sports in the fall *and* spring?

c) How many students play sports *only* in the winter?

6) Use the following information to construct a Venn diagram. 100 students were asked what they do on weekends.

- 24 students only watch television
- 16 students only read
- 12 students only exercise
- 19 students watch television and read
- 14 students watch television and exercise
- 7 students exercise and read
- 8 students do all three activities at some point during each weekend

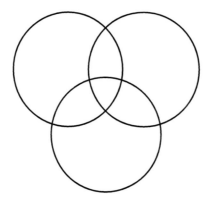

Lesson 55: Mean, Median, and Mode

With a set of data values, it can be useful to have one value that represents the entire data set. **Measures of central tendency** represent centralized values of data.

There are three measures of central tendency:

mean: the average of all the values in a data set

median: the middle value of a data set that is listed in ascending or descending order (Note: If there are an even number of terms in a data set, the median is the average of the middle two terms.)

mode: the value that occurs the most often in a data set

---

example 1

Find the mean, median, and mode for the given set of numbers.

$$\{22, 43, 9, 11, 11, 76, 47, 26, 52\}$$

1) To find the mean, add up each of the numbers and divide by the total number of terms in the data set (9 in this case):

$$\text{mean} = \frac{22+43+9+11+11+76+47+26+52}{9} = \frac{297}{9} = 33$$

2) To find the median, the terms first have to be rearranged in numerical order:

9, 11, 11, 22, 26, 43, 47, 52, 76

Out of nine terms, the median will be the fifth term, 26.

Another approach to find the median involves rearranging the terms in numerical order, and then moving inward from both ends one term at a time:

9 | 11 11 22 26 43 47 52 | 76

Cut off the first and last term, then keep moving towards the middle until there is one term left. That value is the median. (If there are two terms remaining because the data set has an even number of terms, average these two numbers to find the median.)

9 11 | 11 22 26 43 47 | 52 76

9 11 11 | 22 26 43 | 47 52 76

9 11 11 22 | 26 | 43 47 52 76

median = 26

---

3) To find the mode, find the value that appears in the data set more than any other values.

mode = 11

Note: If no values appear more than any of the others, there is **no mode**. If more than one value repeats an equal number of times but more than the rest of the terms, then there are several modes.

---

example 2

Find the mean, median, and mode of the given data set.

$\{5, 6, 6, 7, 8, 8, 9, 10\}$

$$\text{mean} = \frac{5+6+6+7+8+8+9+10}{8} = \frac{59}{8} = 7.375$$

The numbers in this data set are already in numerical order, but there are an even number of terms.

5 6 6 7 8 8 9 10

5 | 6 6 7 8 8 9 | 10

5 6 | 6 7 8 8 | 9 10

5 6 6 | 7 8 | 8 9 10

The middle two numbers are 7 and 8.

$$\text{median} = \frac{7+8}{2} = \frac{15}{2} = 7.5$$

To find the mode, look for which value appears more than any of the others. But both 6 and 8 appear the same number of times and occur more than the other values.

mode = 6, 8

---

example 3

The stem-and-leaf plot below shows the ages of the people who bought skateboards at a store during a sale.

Ages of People

Stem	Leaf
1	1 3 4 5 5 6 6 6 8
2	0 1 7 8
3	9
4	3 6
6	5 5
7	1

Key
6 | 2 = 62

What is the median age of the people who bought skateboards during the sale?

The median is the middle number in a set of numbers that are arranged in order from smallest to largest value, or largest to smallest. A stem-and-leaf plot already displays numbers in ascending or descending order, so finding the median is easy.

In this stem-and-leaf plot, there are 19 ages displayed. The 10th number is the middle value among 19 numbers, so you just need to count up from the smallest value in the plot, or count down from the largest value, until you get to the 10th number, which is 20 in this case.

The median age is 20.

example 4

The graph below shows the frequency of test scores on the algebra final exam.

Results of Algebra Final

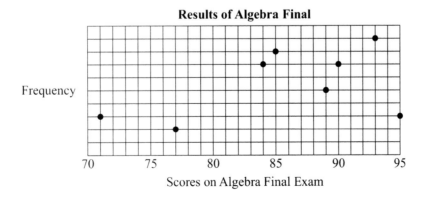

Frequency

70   75   80   85   90   95

Scores on Algebra Final Exam

What is the **mode** of the algebra final exam scores?

A. 88

B. 89

C. 93

D. 95

The mode is the value that occurs most often in a data set, and this graph is displaying how often each test score came up in a list of scores from an exam. Three students scored a 71, two students got a 77 on the exam, and so on. Since we need the score that appeared the most often, we just need to see which score had the largest frequency value. The score of 93 occurred nine times, which is more than the others.

The answer is C.

---

example 5

So far this term, Heidi has these scores on quizzes:

87, 86, 96, 87

What is the lowest score she can get on the one remaining quiz to have a final average (mean) score of 90?

A. 94

B. 97

C. 90

D. 91

After taking this remaining quiz, Heidi will have 5 quiz scores. For all 5 of them to have an average of 90, the 5 scores need to add up to 5 ? 90, or 450 (because the average of any set of numbers comes from adding them all up and dividing by how many numbers you added). This can be expressed as:

$$\frac{87 + 86 + 96 + 87 + x}{5} = \frac{450}{5} = 90$$

Or as: $87 + 86 + 96 + 87 + x = 450$

Solving this expression for x:

$$87 + 86 + 96 + 87 + x = 450$$

$$356 + x = 450$$
$$-356 \qquad -356$$

$$x = 94$$

The answer is A.

Practice Problems

Name _____

Find the mean, median, and mode for each set of data.

1) {42, 33, 50, 46, 37, 48, 52, 43, 55, 39, 37}

2) {310, 540, 820, 700, 650, 490, 770, 380}

3)

stem	leaf
1	3 6 8
2	0 7 7 7
3	2 6
4	4

$1 \mid 3 = 13$

4)

stem	leaf
1	8 8
2	2 5 8 8
3	0 4 7

$1 \mid 8 = 180$

5) What must the next number be for the mean, median, and mode of the set of data to all be the same value?

$$\{14, 17, 18, 18, \underline{\hspace{1cm}} \}$$

6) In the following data set, which is greater, the mean or the median?

$$\{42, 28, 36, 32, 35\}$$

7) The mean of two numbers is 24. What is the sum of these two numbers?

8) If the mean of five consecutive integers is 6, what is the median of the five integers?

9) Gary received the following scores on his first 4 quizzes: 92, 84, 79, 92. What is the lowest grade Gary can get on his next quiz to have an average of at least 87 for the five quizzes?

10) The mean of a, b, c, and d is 22. If the average of a and b is 18, what is average of c and d?

366

Lesson 56: Extremes, Range, and Quartiles

While measures of central tendency represent centralized values of a data set, **measures of variation** are used to represent the spread of a data set.

The measures of variation of a data set are:

lowest value: the smallest numerical value of a data set

greatest value: the largest numerical value of a data set

range: the difference between the greatest and lowest values of a data set

lower quartile: the "median" of all the values less than the median

upper quartile: the "median" of all the values greater than the median

interquartile range: the difference between the upper quartile and the lower quartile

---

example 1

Find the measures of variation of the given set of numbers.

{12, 27, 42, 6, 19, 33, 8, 21, 39, 25, 15}

The first step should always be to rewrite the terms in numerical order:

6 8 12 15 19 21 25 27 33 39 42

Once in order (ascending order is preferred), the lowest and greatest values are easy to identify.

lowest value (LV) = 6

greatest value (GV) = 42

With these two values, the range can now be determined:

range = greatest value – lowest value

= 42 – 6 = 36

Before finding the lower and upper quartiles, the median of the data must be found:

6 8 12 15 19 21 25 27 33 39 42

median = 21

The median divides the data set into two equal parts: values less then the median and values greater than the median. But, the median itself is *not* included in either half.

To find the lower quartile, find the "median" of the values less than the real median:

6 8 **12** 15 19

Of these five values, 12 is the middle term, so the lower quartile = 12.

To find the upper quartile, find the "median" of the values greater than the real median:

25 27 **33** 39 42

Of these five values, 33 is the middle term, so the upper quartile = 33.

Note: The rules for finding the upper and lower quartiles are the same as for finding the median. If there are an even number of terms in the lower and upper half of the data, then each quartile is the average of the middle two terms of the lower or upper half of the data.

Now that we have the values for the upper and lower quartiles, the interquartile range can be found:

interquartile range = upper quartile – lower quartile

$$= 33 - 12 = 21$$

---

example 2

Find the lower quartile and upper quartile for the given data set.

$$\{2, 5, 11, 14, 16, 20, 23, 27, 29\}$$

The median is 16, leaving four values less than the median $\{2, 5, 11, 14\}$ and four values greater than the median $\{20, 23, 27, 29\}$.

For the values less than the median, 5 and 11 are the middle two terms, so the lower quartile is $\dfrac{5+11}{2} = \dfrac{16}{2} = 8$.

For the values greater than the median, 23 and 27 are the middle two terms, so the upper quartile is $\dfrac{23+27}{2} = \dfrac{50}{2} = 25$.

If the median itself is found by averaging together the middle two terms of the entire data set, don't forget to include each of those two values in the lower and upper half of the data when finding the quartiles.

---

example 3

Find the lower quartile and upper quartile for the given data set.

$$\{3, 7, 13, 19, 22, 26, 31, 35, 36, 40\}$$

To find the median, since 22 and 26 are the middle two terms:

$$\text{median} = \frac{22 + 26}{2} = \frac{48}{2} = 24$$

To find each quartile, separate the data into values less than the median and values greater than the median:

3 7 13 19 22   26 31 35 36 40

Notice how, in this example, *every* data value is included because the median isn't one of the actual values from the data set.

lower quartile = 13

upper quartile = 35

---

example 4

The stem-and-leaf plot below shows the number of people using a skateboard park on 13 different days.

Number of Skateboard Park Users

3	0 2
4	2 3 5 6
5	1 4 4 6
6	1 2 4

Key

4 | 3 = 43

What is the range of the data in the stem-and-leaf plot?

A. 29

B. 31

C. 32

D. 34

Remember, one of the great things about a stem-and-leaf plot is that it displays a set of numbers *in order*, from lowest value to greatest value, or from greatest value to lowest value. Either way, when you are asked for the *range*, which is the difference between the greatest value and lowest value, all you need to look at on a stem-and-leaf plot is the first number and last number.

In this stem-and-leaf plot, the first number (and lowest value) is 30 and the last number (and greatest value) is 64. So, to find the difference, or range, subtract 30 from 64:

$$64 - 30 = 34$$

range = 34

The answer is D.

Practice Problems

Name _____

Use the table below to answer questions 1–6.

Average Monthly High and Low Temperatures for Boston (in degrees Fahrenheit)

	Jan	Feb	Mar	Apr	May	Jun	Jul	Aug	Sep	Oct	Nov	Dec
High	36	39	46	56	67	77	82	80	73	62	52	42
Low	22	24	31	41	50	59	65	64	57	46	38	28

1) What is the range of the average high temperatures?

2) What is the range of the average low temperatures?

3) What is the lower quartile of the average high temperatures?

4) What is the upper quartile of the average high temperatures?

5) What is the lower quartile of the average low temperatures?

6) What is the upper quartile of the average low temperatures?

Find the range, median, lower quartile, upper quartile, and interquartile range for each set of data.

7) {6, 3, 7, 5, 4, 2, 6, 3, 8}

8) {28, 37, 17, 21, 25, 12, 30, 15}

9)

stem	leaf
1	2 5 6
2	1 3 3 7
3	5 6
4	1

$1 \mid 2 = 120$

10)

stem	leaf
2	5 9
3	2 6 8
4	1 5 9
5	0 4
6	3

$1 \mid 8 = 180$

Lesson 57: Box-and-Whisker Plots

Some information about the measures of variation and measures of central tendency of a data set can be displayed on a graph called a **box-and-whisker plot**.

Specifically, five values are needed to construct a box-and-whisker plot:

► lowest value

► lower quartile

► median

► upper quartile

► greatest value

Once these five numbers are determined, the box-and-whisker plot is constructed as explained in the following example:

---

example 1

Draw a box-and-whisker plot for the set of data that has the following qualities:

>   lowest value = 3
>   lower quartile = 7
>   median = 12
>   upper quartile = 19
>   greatest value = 27

1) Draw a number line that begins at some value less than the lowest value and ends at some value greater than the greatest value (much the same way a number line is constructed for a line plot).

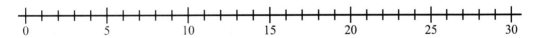

2) Plot the five values as points *above* the number line.

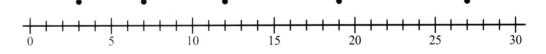

3) Draw a box that ranges from the lower quartile to the upper quartile.

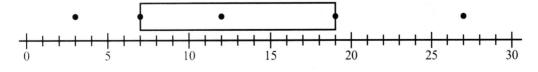

---

4) Draw line segments (the "whiskers") from the lowest value to the lower quartile and from the upper quartile to the greatest value. Also draw a vertical line through the median inside the box you just drew.

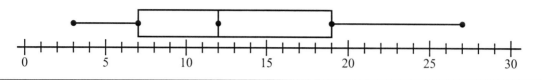

Note: Box-and-whisker plots can be drawn horizontally *or* vertically, although they are usually presented horizontally.

Interpreting Box-and-Whisker Plots

The five values used to draw a box-and-whisker plot (lowest value, lower quartile, median, upper quartile, and greatest value) are the *only* five values that you can determine from a given box-and-whisker diagram. You *cannot* determine the mean or mode.

example 2

Doreen is analyzing a box-and-whisker plot that correctly shows the results of a poll in which a sample of people were asked to state their annual incomes.

Which of the following will Doreen be **unable** to determine from this box-and-whisker plot?

A. the highest annual income among those polled
B. the median annual income among those polled
C. the mean annual income among those polled
D. the lowest annual income among those polled

To answer this question, you need to remember what information a box-and-whisker plot displays. The greatest value of a data set, the lowest value, the median, the upper quartile, and the lower quartile are the only numbers shown on a box-and-whisker plot. You cannot determine the mean or mode.

The answer is C.

A box-and-whisker plot breaks up data into quartiles, or fourths (from Latin, *quarta* means "fourths").

Each "whisker" represents a fourth of the data, while the box represents the middle 50% of the data. The median inside the "box" separates that middle 50% into two 25% portions of the data.

This is shown in the following diagram:

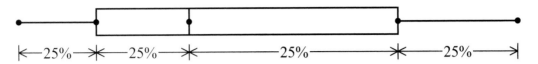

---

example 3

The box-and-whisker plot shown below represents 300 scores on a test given throughout the school district.

How many students scored between 44 and 58?

Since 44 is the lower quartile and 58 is the median, the range of 44 to 58 represents 25% of the student scores. If there are 300 scores, 25% of 300 is 75.

75 students scored between 44 and 58.

---

example 4

The heights of the 20 players on a school soccer team are recorded in the box-and-whisker plot shown below.

Players' Heights in Inches

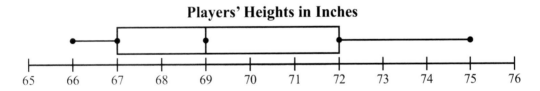

Based on the information given in the box-and-whisker plot, which of the following statements is true?

A. The mean height of the team is 69 inches.
B. Half the players' heights are between 67 and 72 inches.
C. The shortest player on the team is 67 inches.
D. The range of heights of players on the team is 5 inches.

To answer this question, remember that a box-and-whisker plot breaks up data into fourths.

Investigating each answer:

A) The mean of a data set cannot be determined from a box-and-whisker plot.

B) The "box" of this box-and-whisker plot ranges from 67 to 72, so half of these players' heights **are** between 67 and 72 inches.

C) The lowest value on the box-and-whisker plot is 66, not 67.

D) The range of the data is the difference between the greatest value on the box-and-whisker plot (75) and the lowest value (66), which is 9, not 5.

The answer is B, which is the only true statement.

Practice Problems

Name _____

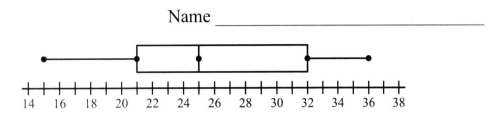

Use the box-and-whisker plot above to answer questions 1–10.

1) What is the median of the data set?

2) What is the upper quartile?

3) What is the lowest value?

4) What is the lower quartile?

5) What is the interquartile range?

6) What is the range?

7) The lowest 25% of the data is less than what value?

8) What percent of the data is represented by the "box"?

9) What percent of the data is less than 32?

10) What percent of the data is represented by each whisker?

Make a box-and-whisker plot from the set of data given in each problem.

11)

stem	leaf
1	4 6 7 8
2	1 3 4 7 9
3	0 4 5
4	2 6 8

1 | 4 = 14

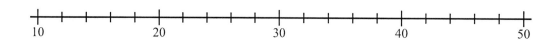

12) {55, 13, 47, 28, 38, 36, 8, 42, 21, 50, 33}

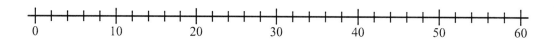

13) {2, 6, 9, 11, 12, 13, 15, 16, 16, 18, 18, 19}

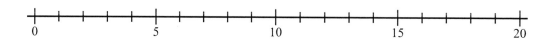

Lesson 58: Data Sampling

Data Sampling

When information is needed about a population, a **survey** is done to collect information from a portion of that population. This portion is called a subgroup or **sample**. For example, surveys you may see on the news about what Americans think about certain issues are done by calling about 1,000 people and asking them, not by calling every single person in the country.

The best kind of sample to use is one chosen at *random*, so that any member of the population has an equal chance of being selected. If you aren't careful to pick members at random, your sample won't be a good representation of the entire population and it will have a **bias**, which means the survey is slanted towards certain members of the population and not all members. For example, if you wanted to conduct a survey of everyone in your town, you wouldn't randomly pick people at a basketball game because you'll have a bias towards people who like basketball. What about everyone else in town that doesn't like basketball?

example 1

A survey is conducted to find the most popular food among ninth-grade students at a school. Which of the following sampling methods would give the **most** accurate results?

A. choose every fourth student in an alphabetical list of all ninth-grade students
B. choose every ninth-grade teacher and every ninth-grade parent
C. choose every fourth student in an alphabetical list of all ninth-grade girls
D. choose every ninth-grade student athlete

Exploring each answer:

Answer A: This answer randomly selects from *all* ninth-grade students by choosing every fourth person, so this answer looks good, but we won't know for sure until we check every possible answer.

Answer B: This answer doesn't allow for any ninth-grade students to be surveyed. Only teachers and parents are being chosen, so this is not a good way to find out what is popular among *students*.

Answer C: This answer randomly selects from ninth-grade *girls* only. To find out what *all* ninth graders like, excluding boys would be a bad idea.

Answer D: This answer excludes any ninth graders that don't play sports, which would not provide an accurate survey of *all* ninth-grade students.

The answer is A.

Types of Sampling

Random Sampling

1) Simple Random Sample

In a **simple random sample**, each member of the population has an equal chance of being chosen.

2) Stratified Sample

In a **stratified sample**, the population is divided into groups (or *strata*), and then some members of each group are randomly selected. (Answer C in **example 1** would be a stratified sample if members of *each* group were randomly selected, males *and* females, and not just the group of female students.)

3) Cluster Sample

In a **cluster sample**, the population is divided into groups and then some of the groups are selected randomly. (Answer D in **example 1** is *almost* a cluster sample, but it chooses only *one* group of students — the student athletes — instead of several groups.)

4) Systematic Sample

In a **systematic sample**, a starting point is selected randomly, and then every n^{th} value of data is taken from a listing of the population. (Answer A in **example 1** is a systematic sample, and Answer C is a systematic sample *of one group* from the population.)

Non-Random Sampling

1) Convenience Sample

A **convenience sample** uses data that is readily available and convenient to collect.

An example of this would be surveying all of your neighbors to get an idea of what the people in your town think about an issue. Your neighbors are a convenient group to ask because they live the closest to you, although your results may not be accurate if people in other neighborhoods feel differently about that issue.

If you aren't sure that your sample is random, one way to help improve the reliability of your survey is to take two samples and then average the results together. This is shown in the example on the next page.

example 2

Four hundred deer were captured in Milltown Forest, tagged, and released back into the forest. Several weeks later, a forest ranger captured a number of deer at a random location in the Milltown Forest, recorded the number of tagged and nontagged deer, and released the deer back into forest. She did this over two trials as shown below.

Record of Deer Captured in Milltown Forest

	Total Number of Deer	Tagged	Nontagged
Trial 1	65	10	55
Trial 2	75	15	60

Approximately how many deer could you expect to find in the entire forest?

A. 2,600

B. 1,600

C. 2,300

D. 1,000

To answer this problem, you first need to determine what ratio to set up. The information provided about the two trials gives the total number of deer captured, the number that were tagged, and the number that were nontagged. So which *two* of these numbers should be used?

The problem states that 400 deer were initially tagged, and you're being asked how many deer are possibly in the entire forest, so use a ratio of tagged deer to the total number of deer, and set up a proportion with the numbers from each trial:

$$\frac{400}{x} = \frac{10}{65} \qquad \text{and} \qquad \frac{400}{x} = \frac{15}{75}$$

Cross-multiply to get:

$$26000 = 10x \qquad \text{and} \qquad 30000 = 15x$$

And then solve each for x:

$$\frac{26000}{10} = \frac{10x}{10} \qquad\qquad \frac{30000}{15} = \frac{15x}{15}$$

$$x = 2600 \qquad \text{and} \qquad x = 2000$$

So which answer is it?

If you set up a proportion with just the numbers from Trial 1, you would think the answer is A, 2600 deer. But with 2 trials, average both results together, which gives a result of 2300.

The answer is C.

THIS PAGE INTENTIONALLY BLANK

Practice Problems

Name _____

Match each data sampling description with the appropriate type of sample.

1) A company has 500 stores across the country. A customer survey is done by randomly choosing 50 of the stores and then randomly selecting 100 customers at each store.

2) A student survey is done by randomly selecting sixth-graders, then randomly selecting seventh-graders, and then randomly selecting eighth graders.

3) Students are surveyed by placing every student name in a box and then names are randomly taken out.

4) A store conducts a survey by talking to the first 20 customers to come into the store one day.

5) A telemarketing company calls every hundredth name in a phone directory.

6) A telemarketing company randomly chooses five states, and then calls random phone numbers in those states.

7) A survey is done selecting every person that walks past a certain street corner.

8) Students are sampled by randomly selecting 25 female students and then randomly selecting 25 male students.

9) A survey is conducted at a football game by selecting every fifth person to walk through the gates.

10) One hundred registered voters are chosen during a survey when a computer randomly selects them from a list of all registered voters.

A) Simple Random Sample

B) Stratified Sample

C) Cluster Sample

D) Systematic Sample

E) Convenience Sample

THIS PAGE INTENTIONALLY BLANK

Lesson 59: Simple Probability

The **probability** of any event occurring is a ratio that describes how likely it is for that event to take place.

Simple probability refers to the probability of one event occurring. (The probability of multiple events occurring, compound probability, will be covered in a later section.)

The ratio that defines simple probability is:

$$\text{probability of an event} = \frac{\text{number of favorable outcomes}}{\text{total number of possible outcomes}}$$

---

example 1

A card is chosen at random from a standard deck of cards. What is the probability of choosing a heart?

There are 52 cards in a standard deck of cards, so there are 52 possible outcomes.

There are 13 hearts in a deck of cards, so there are 13 favorable outcomes.

$$P(\text{event}) = \frac{\text{\# of favorable outcomes}}{\text{total \# of possible outcomes}}$$

$$P(\text{heart}) = \frac{13}{52}$$

This ratio can be reduced to:

$$P(\text{heart}) = \frac{1}{4}$$

The probability of drawing a heart is one out of four.

Any probability ratio can be treated as a regular fraction, which can be turned into a decimal or percent.

$$\frac{1}{4} = 0.25 = 25\%$$

The probability of an event can be a numerical value between 0 and 1, or a percentage between 0 and 100.

This can be expressed as:

$$0 \leq P(E) \leq 1$$

$$0\% \leq P(E) \leq 100\%$$

(where P stands for *probability* and E stands for *event*)

---

example 2

A bag contains 3 blue, 4 red, and 2 white marbles. Karin is going to draw out a marble without looking in the bag. What is the probability that she will **not** draw a red marble?

A. $\dfrac{1}{3}$ C. $\dfrac{2}{3}$

B. $\dfrac{5}{9}$ D. $\dfrac{4}{9}$

There are two ways to approach this problem. 1) Find the probability of drawing any of the marbles that aren't red, or 2) find the probability of drawing a red marble and subtract that from 1 (or from 100%).

Method 1: There are a total of 9 marbles, and 5 of them are **not** red.

$$P(\text{not red}) = \frac{5}{9}$$

Method 2: Out of the 9 marbles, 4 are red:

$$P(\text{red}) = \frac{4}{9}$$

Now subtract this from 1 (remember, 100% = 1):

$$1 - \frac{4}{9} = \frac{9}{9} - \frac{4}{9} = \frac{5}{9}$$

Either method results in the same answer of $\dfrac{5}{9}$.

The answer is B.

---

example 3

Shelley is registering at a hotel that has 14 rooms available on the first floor, 10 rooms available on the second floor, and 16 rooms available on the third floor. If Shelley is assigned one of these hotel rooms at random, what is the probability that it will be on the second floor?

A. $\dfrac{1}{4}$ C. $\dfrac{1}{3}$

B. $\dfrac{3}{10}$ D. $\dfrac{2}{5}$

The total number of rooms from which to choose is 40. The number of rooms on the second floor is 10. So if a room is chosen at random, the probability of it being on the second floor is:

$$P(2^{nd} \text{ floor room}) = \dfrac{10}{40} \text{ or } \dfrac{1}{4}$$

The answer is A.

example 4

The chart below shows the amount spent by customers at a department store on a typical business day.

Amount Spent	Number of Customers
$0	158
$0.01 – $5.99	94
$6.00 – $9.99	203
$10.00 – $19.99	126
$20.00 – $49.99	47
$50.00 – $99.99	38
$100 and over	53

Based on the information in the chart, which of the following is closest to the probability that a customer entering the store on a typical day will spend **at least** $10?

A. 13%

B. 18%

C. 37%

D. 81%

This probability can be found by determining how many customers spent at least $10 and by calculating the total number of customers. The total number of customers can be found by adding up all of the numbers on the right side of the table:

$$158 + 94 + 203 + 126 + 47 + 38 + 53 = 719$$

The number of customers who spent $10 or more can also be added up:

$$126 + 47 + 38 + 53 = 264$$

These two numbers can now be used to calculate the probability of a customer spending at least $10:

$$\text{P(at least \$10)} = \frac{264}{719} = 0.367 = 37\%$$

The answer is C.

---

example 5

What is the probability of spinning an odd number on this spinner?

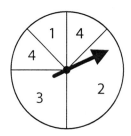

A. $\dfrac{1}{2}$ C. $\dfrac{1}{4}$

B. $\dfrac{3}{8}$ D. $\dfrac{2}{5}$

The first thing to notice about this spinner is that the sections are not all the same size. The probability of spinning a 2, for example, would be higher than spinning any other number. Also, the number 4 is on the spinner twice, although in sections that add up to an area that is still smaller than the section with the 2 in it.

If the spinner is divided into 8 separate sections, all of equal size, it will be easier to see how many of them have an odd number:

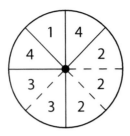

Now that the spinner is divided into equal sections, we can see that the "1" is on 1 section, the "2" is on 3 sections, the "3" is on 2 sections, and the "4" is on 2 sections. So out of all 8 sections of the spinner, an odd number is on 3 of them.

The answer is B.

---

example 6

Chris selected 50 students at random and asked them who they want for class president. The results are shown in the table below.

Candidate	Frequency
Jessica	30
Jeremy	4
Monique	16

Which statement is true regarding the probability that at least 5 of the next 10 students interviewed will want Jeremy for president?

A. It is impossible.
B. It is unlikely.
C. It is likely.
D. It is certain.

To get an idea of how students feel about Jeremy as class president, let's first calculate the probability that any student will select him, based on the results of this survey. Out of the 50 students asked, 4 chose Jeremy. Based on this response rate, the probability that any other student will choose Jeremy is 4 out of 50:

$$P(\text{Jeremy}) = \frac{4}{50} = 0.08 = 8\%$$

Since the probability of any student voting for Jeremy is not zero, it's *not impossible* to run into a group of 10 students, 5 of whom will also want him to be class president. Since the probability is not 100%, it's *not certain* either that at least 5 of the next 10 will want him as class president.

Asking 10 more student who they would like as class president and having at least 5 of them choose Jeremy means running into a group of students where at least 50% of them would vote for him. So far, only 8% are picking him over the other candidates, so it's not likely this would happen. But, as was already pointed out, it's not impossible either. It is, however, unlikely.

The answer is B.

Practice Problems

Name _____

1) When a standard six-sided die is rolled, what is the probability of rolling an even number?

2) When a coin if flipped, what is the probability of it landing "heads" up?

3) The first ten prime numbers are written on a chalkboard. If one is selected at random, what is the probability of choosing a number less than 10? (See page 56 for a list of prime numbers.)

4) A jar contains 7 blue marbles, 4 red marbles, and 9 green marbles. The jar is bumped and one marble pops out. What is the probability of that marble being red?

5) What is the probability of picking an ace out of a standard deck of playing cards?

6) If a day of the week is chosen at random, what is the probability of choosing Wednesday?

7) In the word **PROBABILITY**, if a letter is chosen at random, what is the probability of picking a vowel? (Consider **Y** to be a consonant.)

8) A piggy bank contains 13 pennies, 11 nickels, 19 dimes, and 7 quarters. If a person reaches into the piggy bank without looking, what is the probability of **not** choosing a dime?

Lesson 60: The Fundamental Counting Principle

To determine the number of outcomes possible for multiple events to occur, the
Fundamental Counting Principle can be used.

The Fundamental Counting Principle states that the total number of possible outcomes for
two or more events can be found by multiplying together the number of ways each event
can occur.

example 1

In her closet, Megan has 6 different T-shirts, 5 different pairs of shorts, and 2
different hats. She pulls out 1 T-shirt, 1 pair of shorts, and 1 hat without looking. How
many different combinations of 1 T-shirt, 1 pair of shorts, and 1 hat are possible?

A. 11
B. 16
C. 32
D. 60

Each time Megan pulls out an article of clothing (T-shirt, shorts, or hat), that is a
separate event. A T-shirt can be chosen in 6 different ways, a pair of shorts can be
chosen in 5 different ways, and a hat can be chosen in 2 different ways. The
Fundamental Counting Principle says that the total number of combinations can be
found by multiplying these three numbers together:

$$6 \times 5 \times 2 = 60 \text{ total combinations}$$

The answer is D.

example 2

In how many different ways can the letters in the word SQUARE be arranged?

To create a six-letter word, there are six spaces to fill in and six letters from which to
choose:

____ ____ ____ ____ ____ ____

The first space can be filled with any of the six letters (S, Q, U, A, R, or E). After one
is chosen, there are now five possible letters left for the second space. After choosing
the second letter, there are now four letters remaining for the third space. This
continues until only one letter remains for the last space.

Using the Fundamental Counting Principle, the number of letter arrangements in a six-letter "word" using the letters S, Q, U, A, R, and E is:

$$6 \times 5 \times 4 \times 3 \times 2 \times 1 = 720$$

There are 720 different ways to arrange the letters in the word SQUARE.

You can see that the more items there are, the number of ways to arrange those items gets large in a hurry.

Some problems, which look similar to the previous examples, can **not** be solved by using the Fundamental Counting Principle.

---

example 3

Six students are participating in a fitness program. They are required to work out in pairs. How many **different** combinations of pairs of students are possible?

A. 3

B. 5

C. 15

D. 30

Let's call the six students A, B, C, D, E, and F. If all of the possible combinations are written out, the results are:

AB	BC	CD	DE	EF
AC	BD	CE	DF	
AD	BE	CF		
AE	BF			
AF				

Note: BA is not written in the second column because it is the same as AB, which has already been counted.

Adding up all of the possible pairs instead of multiplying gives a total of 15.

The answer is C.

Practice Problems

Name _____

1) How many different "words" can be made with the letters **M**, **A**, **T**, and **H**?

2) When two six-sided dice are rolled together, how many possible outcomes are there?

3) A customer at a restaurant buys the lunch special, which includes a sandwich, a bowl of soup, and a drink. If the customer can choose from 3 types of sandwiches, 2 different kinds of soup, and either soda or iced tea, how many lunch special combinations are possible?

4) The Environmental Club has 8 sixth-graders, 11 seventh-graders, and 15 eighth graders. If one person from each grade is chosen to go to an event in Boston, how many combinations of students are possible?

5) Alexis is getting dressed one morning. She can choose from 4 pairs of shorts, 6 shirts, and 2 pairs of sneakers. How many different outfits can she make with those items?

6) If there are 3 airlines that fly between Boston and Baltimore, and 4 airlines that fly between Baltimore and Los Angeles, how many different ways can you fly from Boston to Los Angeles by going through Baltimore?

7) Because of cell phones, more area codes are being added around the country. How many area codes are possible? (The first digit cannot be a 0 or a 1.)

8) License plates issued in a particular state have 4 digits (0–9) followed by 2 letters. If the digits and letters can be repeated, how many different license plates are possible with 4 digits and 2 letters?

Lesson 61: Compound Probability

A **compound event** consists of two or more single (or simple) events.

Compound probability involves finding the probability of a compound event occurring in a certain way.

The definition of probability is still

$$P(\text{event}) = \frac{\text{\# of favorable outcomes}}{\text{total \# of possible outcomes}}$$

but the Fundamental Counting Principle may be needed to determine the number of favorable outcomes or total number of possible outcomes for a compound event.

Independent Events

When two or more events have outcomes that have nothing to do with each other, they are called **independent events**. When a compound event involves two or more independent events, the probability of that compound event occurring can be found by multiplying together the simple probability of each independent event occurring.

---

example 1

60% of the cars owned by Best Car Rental are white and 30% have a standard transmission. If you randomly choose a rental car, what is the probability that you will get a white car **with** a standard transmission?

A. $\dfrac{9}{10}$ C. $\dfrac{9}{100}$

B. $\dfrac{18}{100}$ D. $\dfrac{90}{100}$

Choosing the color of the car and choosing the transmission type can be considered independent events. So, to find the probability of the compound event, multiply together the probabilities of each independent event.

$60\% \times 30\% =$

$0.60 \times 0.30 =$

$\dfrac{6}{10} \times \dfrac{3}{10} = \dfrac{18}{100}$

The answer is C.

---

example 2

When a die is rolled and a coin is flipped, what is the probability of rolling a 4 on the die **and** getting "tails" on the coin?

Flipping a coin and rolling a die are independent events. The outcome of the coin couldn't possibly affect the outcome of the die. So, to find the probability of this compound event occurring, find the probability of each event separately and then multiply the two simple probabilities together.

<u>coin</u>   <u>die</u>

$$P(\text{tails}) = \frac{1}{2} \qquad\qquad\qquad P(4) = \frac{1}{6}$$

$$P(\text{tails and 4}) = \frac{1}{2} \times \frac{1}{6} = \frac{1}{12}$$

To prove this answer, a **tree diagram** can be drawn to show all of the possible outcomes of this compound event.

To draw a tree diagram:

1) Vertically list all of the possible outcomes of the first event.

T (tails)

H (heads)

2) Next to *each* of these outcomes, vertically list all of the possible outcomes of the second event and draw line segments from each of the first outcomes to each of the second outcomes.

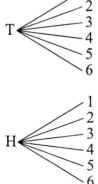

This is called a "tree diagram" because each of the second possible outcomes is shown "branching" away from each of the first possible outcomes.

A tree diagram can be used to list all of the possible outcomes of the compound event. (A list of all possible outcomes for an event is called a **sample space**.)

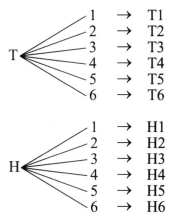

$$
\begin{array}{ccc}
1 & \rightarrow & T1 \\
2 & \rightarrow & T2 \\
3 & \rightarrow & T3 \\
T \quad 4 & \rightarrow & T4 \\
5 & \rightarrow & T5 \\
6 & \rightarrow & T6 \\
\end{array}
$$

$$
\begin{array}{ccc}
1 & \rightarrow & H1 \\
2 & \rightarrow & H2 \\
3 & \rightarrow & H3 \\
H \quad 4 & \rightarrow & H4 \\
5 & \rightarrow & H5 \\
6 & \rightarrow & H6 \\
\end{array}
$$

There are twelve possible outcomes in the sample space, and only one shows "tails" for the coin and "4" for the die (T4). Therefore, the probability is 1 out of 12.

---

example 3

What is the probability of drawing two cards simultaneously from a standard deck and having them both be aces?

Even though the cards are being drawn together, we need to treat the drawing of each card as a separate event. There are four aces in a standard deck of fifty-two cards, so the probability of the first card being an ace is 4 out of 52, which can be reduced to 1 out of 13.

If the first card drawn was an ace (and we must assume that it was when calculating probability), that would mean there are now three aces left in a deck that has fifty-one cards left. So the probability of the second card being an ace is 3 out of 51. The outcome of this second event was affected by the first event because there are less aces from which to choose and less cards in the deck now. However, by assuming that the first card was an ace, we can treat these events at if they were independent.

Once you have the two probabilities, you can find the overall probability of this compound event by multiplying the probabilities together:

$$
\frac{1}{13} \times \frac{3}{51} = \frac{3}{663} = 0.0045
$$

The probability of drawing two aces from a deck is 3 out of 663, or 0.45%.

---

example 4

A bag contains 2 blue, 6 black, and 4 white socks. Paula is going to draw out a sock without looking in the bag. What is the probability that she will draw **either** a blue **or** black sock?

A. $\dfrac{1}{6}$ C. $\dfrac{1}{3}$

B. $\dfrac{1}{2}$ D. $\dfrac{2}{3}$

When finding a probability where *either* of two possible outcomes is acceptable (the answer can be one outcome *or* a different outcome), instead of finding the probability of each event and multiplying them together, *add* the two probabilities.

The probability of each event needs to be found:

probability of selecting a blue sock: $\dfrac{2}{12}$ or $\dfrac{1}{6}$ because 2 of the 12 socks are blue

probability of selecting a black sock: $\dfrac{6}{12}$ or $\dfrac{1}{2}$ because 6 of the 12 socks are blue

Now add these two probabilities together:

$$\frac{1}{6} + \frac{1}{2} = \frac{2}{12} + \frac{6}{12} = \frac{8}{12} \text{ or } \frac{2}{3}$$

Another way to see where this answer comes from is to realize that 8 of the 12 socks are either blue or black. The answer is D.

(Note: This problem was an example of simple probability, as only one event occurred, but one way the answer could be found involved treating each acceptable outcome as a separate event.)

Dependent Events

When the outcome of one event affects the outcome of another event, they are called **dependent events**. Tree diagrams can be used for solving some compound probability problems that involve dependent events.

---

example 5

If a couple has three children, what is the probability that two of them are girls?

To create the sample space for this problem, make a tree diagram. The first child can be a boy or a girl, the second child can be a boy or a girl, and the third child can be a boy or a girl:

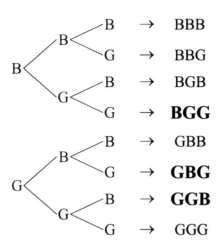

There are eight possible outcomes for the genders of the three children, and three of those outcomes show two girls and one boy. So the probability of a couple having three children, two of which are girls, is 3 out of 8.

Some compound probability problems involving dependent events require other ways of determining the number of favorable outcomes or possible outcomes.

---

example 6

Use the spinners to answer the question.

When playing a board game, you spin two spinners with congruent sectors numbered 1 through 7 as shown. If the sum of the two numbers you spin is 12, 13, or 14, you win. What is the probability of winning?

A. $\dfrac{21}{49}$ C. $\dfrac{15}{49}$

B. $\dfrac{10}{49}$ D. $\dfrac{6}{49}$

To determine the number of favorable outcomes, the number of ways a 12, 13, or 14 can occur needs to be calculated.

12		13		14	
spinner 1	spinner 2	spinner 1	spinner 2	spinner 1	spinner 2
5	7	6	7	7	7
6	6	7	6		
7	5				

There are 6 ways to get a 12, 13, or 14.

To determine the total number of possible outcomes, the Fundamental Counting Principle can be used. There are seven possible outcomes for the first spinner and seven possible outcomes for the second spinner.

$$7 \times 7 = 49$$

There are 49 possible outcomes for the two spinners, therefore:

$$P(\text{win}) = \frac{6}{49}$$

The answer is D.

Compound Probability Involving Two Dice

Some questions about compound probability may involve two dice.

example 7

If two dice are tossed, what is the probability that the sum of the two dice is 6?

Instead of a tree diagram, the following chart can be constructed:

Die #1

		1	2	3	4	5	6
	1	2	3	4	5	6	7
	2	3	4	5	6	7	8
Die #2	**3**	4	5	6	7	8	9
	4	5	6	7	8	9	10
	5	6	7	8	9	10	11
	6	7	8	9	10	11	12

The numbers across the top are the possible outcomes of the first die, and the numbers down the left side are the possible outcomes of the second die. All the numbers inside the chart are the sums of the two outcomes of the dice.

When two dice are rolled, there are 36 total possible outcomes and 5 of them are a sum of 6 from the two dice. So, the probability of rolling a total of 6 with two dice is 5 out of 36, or 13.9%.

Practice Problems

Name _____

1) A coin is flipped and a six-sided die is rolled. What is the probability of flipping "tails" on the coin and rolling a 3 on the die?

2) There are 4 questions on a true-or-false quiz. If a student guessed on all 4 questions, what is the probability of getting them all right?

3) A jar contains 10 cookies: 3 chocolate chip cookies, 2 sugar cookies, and 5 oatmeal cookies. Mike randomly takes out one cookie and eats it. He then takes another cookie at random. What is the probability that both cookies were chocolate chip?

4) If two standard dice are rolled, what is the probability of rolling doubles (the same number on each die)?

5) Two cards are drawn from a standard deck of playing cards. One card is removed and then placed back into the deck before the second card is drawn. What is the probability of drawing an ace both times?

6) Two cards are drawn from a standard deck of playing cards. One card is removed and **not** placed back into the deck before the second card is drawn. What is the probability of drawing an ace both times?

7) A family has three children. What is the probability that the family has *at least* two girls?

8) A coin is flipped three times. What is the probability that "tails" comes up exactly twice?

Index

Answer Key

Unit 1 – Number Sense

Lesson 1

1) F – Additive Identity Property
2) C – Associative Property of Addition
3) I – Multiplicative Inverse Property
4) H – Additive Inverse Property
5) D – Associative Property of Multiplication
6) A – Commutative Property of Addition
7) J – Multiplicative Property of Zero
8) I – Multiplicative Inverse Property
9) A – Commutative Property of Addition
10) G – Multiplicative Identity Property
11) D – Associative Property of Multiplication
12) B – Commutative Property of Multiplication
13) E – Distributive Property
14) B – Commutative Property of Multiplication
15) C – Associative Property of Addition
16) F – Additive Identity Property
17) H – Additive Inverse Property
18) J – Multiplicative Property of Zero
19) G – Multiplicative Identity Property
20) E – Distributive Property
21) $2a + 2b$
22) $-12m + 8n$
23) $x - y - 4$
24) $-2m - 4$
25) $ab - 8b$
26) $x + 2y$
27) $-2k^2 - 5k$
28) $5a + 8$
29) $\dfrac{1}{9}$
30) 3
31) $\dfrac{x}{2}$
32) $\dfrac{1}{5m}$
33) $\dfrac{3}{a + 2}$
34) $\dfrac{4}{9}$

Lesson 2

1) 3.76
2) 17.253
3) 3.57
4) -3.05
5) 23.014
6) 28.152
7) 3.7
8) 15
9) 14.54, 14.94, 15.45, 15.55
10) 38, 38.9, 39, 39.8
11) 1,000,001; 1,000,010; 1,000,100; 1,001,000
12) 14.8, 14.889, 14.89, 14.9
13) rational
14) rational
15) irrational
16) rational
17) rational
18) irrational
19) rational
20) rational

Lesson 3

1) $2\sqrt{5}$
2) -8
3) 30
4) 0.4
5) -0.05
6) $\dfrac{4}{5}$
7) $\dfrac{1}{3}$
8) $\dfrac{2\sqrt{3}}{3}$
9) $\dfrac{3\sqrt{7}}{7}$
10) 16
11) $70\sqrt{2}$
12) -6
13) 3 and 4
14) 5 and 6
15) 3 and 4
16) 7 and 8
17) 2 and 3
18) 1 and 2
19) true
20) false
21) true
22) false

Lesson 4

1) 4.5×10^3
2) 7.71×10^4
3) 6×10^9
4) 5.9×10^{-3}
5) 4.82×10^{-4}
6) 2.3×10^{-2}
7) 3.04×10^{-3}
8) 1.93×10^6
9) 4.3×10^{-6}
10) 37,400,000
11) 466
12) 8,000,000
13) 0.00064
14) 0.0945
15) 0.0000025
16) 190,000
17) 256,000
18) 0.473
19) 4.5×10^5; 450,000
20) 3.834×10^7; 38,340,000
21) 5.4×10^{-6}; 0.0000054
22) 3.75×10^{-6}; 0.00000375
23) 6×10^{-5}; 0.00006
24) 2×10^3; 2000
25) 3×10^3; 3000
26) 2×10^6; 2,000,000

Lesson 5

1) $\dfrac{5}{9}$
2) $\dfrac{3}{8}$
3) $\dfrac{2}{5}$
4) $\dfrac{5}{8}$
5) $\dfrac{19}{24}$
6) $\dfrac{29}{35}$
7) $\dfrac{1}{6}$
8) $\dfrac{1}{6}$
9) $\dfrac{3}{20}$

10) $\dfrac{1}{9}$
11) $\dfrac{7}{15}$
12) $\dfrac{9}{22}$
13) $\dfrac{5}{2}$
14) $\dfrac{4}{3}$
15) $\dfrac{6}{5}$
16) <
17) =
18) >
19) $\dfrac{2}{5}, \dfrac{3}{7}, \dfrac{4}{9}$
20) $\dfrac{2}{9}, \dfrac{3}{13}, \dfrac{1}{4}$
21) $\dfrac{4}{7}, \dfrac{3}{5}, \dfrac{5}{8}$

Lesson 6

1) 64
2) 0
3) 2
4) 6
5) 33
6) 9
7) 16
8) 3
9) 32
10) 2
11) 9
12) $\sqrt{13}$
13) 1
14) 38
15) 25
16) 2
17) 8
18) 3
19) 7
20) 2

Lesson 7

1) 43%
2) 76%
3) 101%
4) 0.2%
5) 40%
6) 311%
7) 570%
8) 123.4%
9) 400%
10) 0.75
11) 0.09
12) 1
13) 0.5
14) 0.065
15) 0.333
16) 10.01
17) 0.005
18) 0.46
19) 40%
20) 120%
21) 35%
22) $16.\overline{6}\%$
23) 62.5%
24) 75%
25) $44.\overline{4}\%$
26) 70%
27) $61.\overline{538461}\%$
28) $\dfrac{4}{25}$
29) $\dfrac{1}{2}$
30) $\dfrac{3}{2}$
31) $\dfrac{1}{3}$
32) $\dfrac{2}{5}$
33) $\dfrac{1}{25}$
34) $\dfrac{1}{8}$
35) $\dfrac{1}{20}$
36) $\dfrac{3}{4}$
37) 5

38) 16
39) 25%
40) 40
41) 50%
42) 120%

Lesson 8

1) −25%
2) −10%
3) $33.\overline{3}\%$
4) **150%**
5) −50%
6) 20%
7) 16
8) 100
9) 100
10) 80
11) 200
12) 60
13) 192
14) 25
15) 49
16) 44
17) 198
18) 90
19) 78
20) 100
21) 20
22) $110
23) $10.49

Lesson 9

1) $2 \cdot 3 \cdot 5$
2) $2 \cdot 2 \cdot 3 \cdot 3 \cdot 3$
3) $5 \cdot 5 \cdot 7$
4) $2 \cdot 3 \cdot 5 \cdot 7$
5) $2 \cdot 2 \cdot 2 \cdot 2 \cdot 2 \cdot 13$
6) $2 \cdot 3 \cdot 3 \cdot 5 \cdot 7$
7) $5 \cdot 5 \cdot 5 \cdot 7$
8) $2 \cdot 3 \cdot 5 \cdot 7 \cdot 11$
9) 10
10) 8
11) 12
12) 27
13) 128
14) 8
15) 4
16) 9

Lesson 10
1) $x = 1$
2) $a = -9$
3) $p = -42$
4) $d = 6$
5) $b = \dfrac{10}{3}$
6) $n = -\dfrac{12}{7}$
7) $x = 7$
8) $c = 13$
9) $h = -6$
10) $m = 1$
11) 400 female students
12) 5 bags
13) 15 lawns
14) 72 houses
15) 360 pieces of fruit
16) 300 maple trees

Lesson 11
1) $y = 25$
2) $x = -6$
3) $y = -12$
4) $x = -12$
5) $y = -27$
6) $x = 5$
7) $y = -\dfrac{84}{9}$
8) $x = 9$
9) $y = 8$
10) $x = 4$
11) $y = -4$
12) $x = -10$
13) $y = -2$
14) $x = 4$
15) $y = -\dfrac{10}{3}$
16) $x = 27$
17) $42.00
18) 20 packages
19) $6900
20) $10.50
21) 6 hours
22) 10 feet
23) 6 hours
24) 15 minutes

Lesson 12
1) 30, 36
2) 64, 128, 256
3) 15, 10, 5, 0
4) 250, -125
5) $a_{12} = 74$
6) $a_7 = 4096$
7) $a_{30} = 107$
8) $a_{14} = 8192$
9) $a_6 = 33$
10) $a_{12} = 6144$
11) $a_8 = 33$
12) $a_{11} = \dfrac{3}{4}$ or 0.75
13) $a_{14} = 22$
14) $a_6 = \dfrac{2}{125}$ or 0.016
15) $a_n = 8 + (n - 1) \times 6$
16) $a_n = 4(3)^{n-1}$
17) $a_n = 85 + (n - 1) \times (-5)$
18) $a_n = 1944\left(\dfrac{1}{3}\right)^{n-1}$

Lesson 13
1) 89, 144
2) 68, 110
3) 123, 199
4) 39, 48
5) 9, 2
6) 21, 20
7) 22, 24
8) 141, 205
9) 59, 78
10) 10, 19
11) D.
12) C.

Lesson 14
1) 1 8 28 56 70 56 28 8 1
2) 78 squares
3) to the right, ↳
4) CIRCLE
5) ①
6) $5x^5, 6x^6$
7) ●, □
8) χ, δ
9) $a + 36, a + 49$
10) V5, U6

Lesson 15
1) not linear
2) not linear
3) linear
4) linear
5) not linear
6) linear
7) not linear
8) linear
9) linear
10) not linear
11) not linear
12) linear
13) $y = -3x + 8$
14) $y = 3x - 2$
15) $y = 2x + 5$
16) $y = -\dfrac{1}{2}x + 2$

Lesson 16
1) $m = 4$
2) $m = 0$
3) $m = \dfrac{3}{4}$
4) no slope
5) $m = 2$
6) $m = 0$
7) $m = -2$
8) $m = \dfrac{5}{6}$
9) $c = 5$
10) $c = 8$
11) $c = 8$
12) $c = 3$
13) $c = 2$
14) $c = 5$

Lesson 17
1) $y - 2 = 5(x - 4)$
2) $y - 0 = -\dfrac{2}{3}(x - 0)$ or $y = -\dfrac{2}{3}x$
3) $y + 5 = 3(x - 0)$ or $y + 5 = 3x$
4) $y + 1 = \dfrac{4}{5}(x + 3)$
5) $y + 4 = \dfrac{1}{2}(x - 2)$ or $y + 5 = \dfrac{1}{2}(x - 0)$
6) $y - 1 = -\dfrac{2}{3}(x - 5)$ or $y - 3 = -\dfrac{2}{3}(x - 2)$
7) $y - 2 = 1(x - 3)$ or $y - 0 = 1(x - 1)$

8) $y - 4 = -1(x - 0)$ or $y - 6 = -1(x + 2)$
9) $m = \dfrac{2}{3}$; y-int. $= (0, -4)$
10) $m = -\dfrac{1}{2}$; y-int. $= (0, 5)$
11) $2x - 3y = 18$
12) $12x + 4y = 1$
13) $y = \dfrac{1}{2}x - \dfrac{3}{4}$
14) $y = 4x - 2$
15) $y = \dfrac{1}{3}x - \dfrac{8}{3}$
16) $y = -\dfrac{5}{2}x + 5$
17) $3x - y = -7$
18) $2x + 3y = -11$
19) $y + 4 = \dfrac{1}{3}(x - 3)$ or $y + 5 = \dfrac{1}{3}(x - 0)$
20) $y + 1 = -2(x - 4)$ or $y - 3 = -2(x - 2)$
21) $y = -\dfrac{3}{7}x + \dfrac{44}{7}$
22) $y = -3$
23) $y = 4$
24) $3x + 2y = 15$

Lesson 18
1)
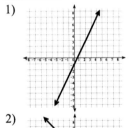
2)
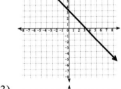
3)
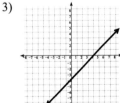

4)
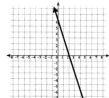

5)
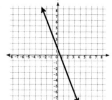

6)
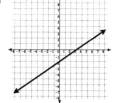

7)
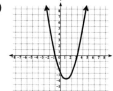

8)
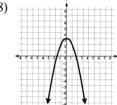

Lesson 19
1) $y = 2x - 5$

2) $y = -\dfrac{1}{3}x - \dfrac{4}{3}$

3) $y = -3x - 19$

4) $y = -4x + 12$

5) $y = 4$

6) $y = \dfrac{5}{4}x + \dfrac{1}{4}$

7) $y = \dfrac{2}{7}x - \dfrac{8}{7}$

8) $x = 6$

9) $y = -\dfrac{5}{2}x - 1$

10) $y = 0$

11) $y = \dfrac{1}{4}x + \dfrac{15}{4}$

12) $y = \dfrac{1}{5}x - \dfrac{18}{5}$

13) $y = \dfrac{2}{3}x + \dfrac{19}{3}$

14) $y = -5x - 18$

15) $x = 0$

16) $y = \dfrac{4}{7}x + \dfrac{40}{7}$

Lesson 20
1) $b = 14$

2) $x = 0$

3) $d = 22$

4) $a = \dfrac{1}{3}$

5) $f = 10$

6) $m = \dfrac{19}{15}$

7) $n = 8$

8) $x = 6$

9) $b = 35$

10) $d = -7$

11) $h = -16$

12) $g = -\dfrac{2}{15}$

13) $x = 7$

14) $m = 4$

15) $c = 8$

16) $g = 7$

17) $d = 5$

18) $x = 5$

19) $x = 12$

20) $a = 0$

21) $m = -27$

22) $t = 12$

23) $a = 7$

24) $p = -24$

25) 27, 29, 31

26) 6, 8

27) 7 inches × 11 inches

Lesson 21
1) $b < 9$

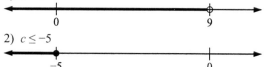

2) $c \le -5$

3) $a \leq -7$

4) $y \geq 8$

5) $n > 30$

6) $p < -\dfrac{11}{2}$

7) $d < 10$

8) $t \geq -7$

9) $n > 2$

10) $z < -15$

11) $s > -2$

12) $b < -1$

13) $n \leq 16$

14) $n \leq 12$

15) $n > -6$

Lesson 22

1) $x = 3$ or 11

2) $c = -2$ or 10

3) $x = \dfrac{1}{3}$ or $\dfrac{11}{3}$

4) $b = -2$ or 3

5) no solution

6) $p = -\dfrac{3}{4}$ or $\dfrac{11}{12}$

7) $a = 0$ or 6

8) $x = -12$ or 8

9) $t = -\dfrac{21}{9}$ or $\dfrac{29}{9}$

10) $x = -\dfrac{26}{7}$ or $\dfrac{30}{7}$

11) $g = -\dfrac{232}{15}$ or $\dfrac{248}{15}$

12) $k = -\dfrac{12}{21}$ or $-\dfrac{12}{27}$

13) $x < -8$ or $x > 2$

14) $m > -1$ and $m < 5$

15) $x \leq -6$ or $x \geq 6$

16) $x \geq 1$ and $x \leq 9$

17) $d > 2$ and $d < 4$

18) $z \leq -5$ or $z \geq -3$

19) $f < -\dfrac{15}{2}$ or $f > \dfrac{9}{2}$

20) $c \geq -\dfrac{1}{12}$ and $c \leq \dfrac{17}{12}$

21) $x =$ all real numbers

22) $f > -\dfrac{45}{20}$ and $f < \dfrac{45}{20}$

Lesson 23

1) $x = \dfrac{y + 5}{2}$

2) $x = \dfrac{h - 3}{-a}$ or $x = \dfrac{3 - h}{a}$

3) $x = \dfrac{m + f}{b}$

4) $x = 9a + 12$

5) $x = \dfrac{2b + 5}{4}$

6) $x = \dfrac{4d - 3c}{6}$

7) $x = \dfrac{4dy - h}{-2}$ or $x = \dfrac{h - 4dy}{2}$

8) $x = y - mn$

9) $x = \dfrac{b}{a + 4}$

10) $m = \dfrac{F}{a}$

11) $n = \dfrac{pV}{RT}$

12) $f = \dfrac{1}{T}$

13) $v = \sqrt{\dfrac{2K}{m}}$

14) $v = 36$

Lesson 24
1) a^5
2) $8x^5$
3) $m^4 n^7$
4) c^9
5) $9a^6$
6) $36x^2 y^2$
7) $\dfrac{m^4}{n^4}$
8) $108x^{12}$
9) $a^5 b^5$
10) $x^6 y^7$
11) $a^5 b^5 c^2$
12) $-2e^{14} f^{13}$
13) $-27m^8 n^6$
14) $\dfrac{a^9}{b^{12}}$
15) $-128x^8 y$
16) a^2
17) xy^4
18) $\dfrac{m^2}{3n^2}$
19) $\dfrac{1}{b^3}$
20) $m^6 n^4$
21) $\dfrac{x^3}{y^5}$
22) $-\dfrac{1}{3c^3 d}$
23) $\dfrac{1}{2a^2 b^3}$
24) $4m^2 n^5$
25) $\dfrac{16a^4}{9b^6}$
26) $x^5 y^3$
27) $\dfrac{3mp^2}{n^4}$
28) $\dfrac{9a}{8b^4}$
29) $\dfrac{3}{x^3 y}$
30) $3x^2 y^2$

Lesson 25
1) $6x^2 - 14x$
2) $-30x^5 - 10x^3 y^2$
3) $a^4 b^5 - 3a^3 b^6$
4) $\dfrac{2}{15} m^4 n^4 - \dfrac{4}{9} m^3 n^5$
5) $10b^4 - 20b^3 + 15b^2$
6) $8x^6 y^2 - 2x^5 y^3 + 6x^4 y^4$
7) $x^2 + 3x - 18$
8) $2y^2 + 11y - 21$
9) $4a^2 - 7ab - 2b^2$
10) $3m^3 - 12m^2 n + mn - 4n^2$
11) $x^3 - 7x + 6$
12) $18c^3 + 3c^2 - 43c + 22$
13) $m^2 + 2mn + n^2$
14) $4p^2 + 12pq + 9p^2$
15) $9c^2 - 6cd + d^2$
16) $25g^2 - 20gh + 4h^2$
17) $x^2 - y^2$
18) $9a^2 - 16b^2$
19) $4m^4 - 1$
20) $\dfrac{1}{4} x^2 - \dfrac{1}{9} y^2$

Lesson 26
1) $3a(a - 4)$
2) $-5y(5y^2 - 2)$
3) $ab(2a + 1)$
4) $2x(x^2 + 6x - 4)$
5) $7c(2c^2 d^2 - 3cd + 5)$
6) $g^2(5g^2 - 7g + 1)$
7) $2x(2x^2 + 5y^2 - x + 3)$
8) $(m + 5)(m + 2)$
9) $(a + 2)(a - 3)$
10) $(b + 2)(b + 6)$
11) $(m + 6)(m - 3)$
12) $(t - 3)(t - 4)$
13) $(g + 11)(g - 4)$
14) $(d + 13)(d + 4)$
15) $(3x - 2)(5x - 2)$
16) $(2y + 1)(3y - 4)$
17) $(3p + 2)(2p + 1)$

18) $(2x - 3y)^2$

19) $(m + 4n)^2$

20) $(5c - 2d)^2$

21) $2(z + 5)(z - 4)$

22) $4(x + 2)(x + 1)$

23) $3(2x - 1)(x + 3)$

24) $(x + y)(x - y)$

25) $(c + 2d)(c - 2d)$

26) $4(m + 3n)(m - 3n)$

27) $3(3m + 2n)(3m - 2n)$

Lesson 27

1) $x = -4, 0$

2) $c = 0, 5$

3) $b = -2, 3$

4) $a = -\dfrac{3}{2}, \dfrac{1}{3}$

5) $c = -\dfrac{4}{3}$

6) $f = -10$

7) $h = 0, 5$

8) $b = 0, 2$

9) $x = -4, -1$

10) $y = -3, 4$

11) $a = -\dfrac{1}{3}, 4$

12) $b = -\dfrac{3}{2}, \dfrac{5}{4}$

13) $e = -2, 0, 7$

14) $m = -\dfrac{3}{2}, \dfrac{3}{2}$

15) $x = -4, -\dfrac{1}{2}$

16) $m = \dfrac{-3 \pm \sqrt{17}}{2}$

17) $b = -2, \dfrac{1}{3}$

18) $n = \dfrac{7 \pm \sqrt{37}}{2}$

19) $a = \dfrac{-1 \pm \sqrt{10}}{3}$

20) $y = \dfrac{5 \pm \sqrt{17}}{2}$

Lesson 28

1) not a function

2) function

3) not a function

4) function

5) not a function

6) function

7) function

8) not a function

9) $f(3) = 10$

10) $g(5) = 18$

11) $h(-2) = -15$

12) $f(-5) = -14$

13) $g(0) = 3$

14) $h(4) = 57$

15) $f(13) = 40$

16) $g\left(\dfrac{1}{2}\right) = \dfrac{9}{4}$

17) $h(-1) = -8$

18) $f(y) = 3y + 1$

19) $g(x + 2) = x^2 + 2x + 3$

20) $h(a) = a^3 - 7$

Lesson 29

1)

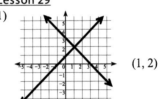

$(1, 2)$

2)

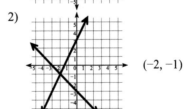

$(-2, -1)$

3)

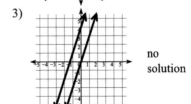

no solution

4)

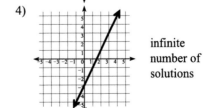

infinite number of solutions

5) infinite number of solutions
6) (3, 5)
7) (−1, 3)
8) no solution
9) (6, 1)
10) (−2, 0)
11) no solution
12) (−3, 0)
13) (4, −1)
14) (1, 1)
15) (−4, −5)
16) infinite number of solutions

Lesson 30
1) 20
2) 60 shirts
3) $13.45
4) 24 students
5) 8 and 10
6) 3 × 7
7) 960 students
8) Samantha is 13 years old;
her brother is 7 years old
9) 300 sophomores
10) $79.00

Lesson 31
1) 6 quarters and 10 dimes
2) 2 liters
3) 1 quart
4) 30 miles per hour
5) 23 cups
6) 4 hours
7) 10 miles per hour
8) $1\frac{1}{3}$ kilograms

Unit 3 – Geometry and Measurement

Lesson 32
1) 135º
2) 30º
3) 90º
4) 60º
5) 45º
6) 30º
7) 90º
8) 180º
9) ∠A and ∠C
10) ∠A and ∠B; ∠A and ∠D; ∠C and ∠D; ∠B and ∠C

11) ∠A and ∠C
12) ∠B and ∠D
13) comp: 75º; supp: 165º
14) comp: no complement; supp: 90º
15) comp: no complement; supp: no supplement
16) comp: 60º; supp: 150º
17) comp: 45º; supp: 135º
18) comp: no complement; supp: 85º
19) comp: no complement; supp: 45º
20) comp: no complement; supp: no supplement
21) 32º and 58º
22) 60º and 120º

Lesson 33
1) 60º
2) 120º
3) 60º
4) 120º
5) 60º
6) 120º
7) ∠1, ∠3, ∠5, ∠7
8) ∠2, ∠4, ∠6, ∠8
9) 120º
10) 240º

Lesson 34
1) 70º
2) 90º
3) 60º
4) 45º
5) 90º
6) 90º
7) 40º
8) 1º
9) $(120 − x)$º
10) $(170 − 2a)$º
11) 130º
12) 50º
13) 40º
14) 140º
15) 90º
16) 90º
17) 35º, 85º, 60º
18) 40º, 50º, 90º
19) 90º, 70º, 20º
20) 60º, 60º, 60º
21) 40º
22) 50º
23) 100º
24) 80º

Lesson 35
1) $c = 5$
2) $b = 12$
3) $a = 1$
4) $c = \sqrt{13}$
5) yes
6) no
7) no
8) yes
9) yes
10) no
11) no
12) yes
13) $\sin A = \dfrac{3}{5}$ or 0.6; $\cos A = \dfrac{4}{5}$ or 0.8; $\tan A = \dfrac{3}{4}$ or 0.75
14) 50°
15) 70°
16) 61°
17) 26°
18) 37°
19) 6.93
20) 4.24

Lesson 36
1) 540°
2) 720°
3) 1080°
4) 3420°
5) 360°
6) 900°
7) 10 sides
8) 6 sides
9) 8 sides
10) 5 sides
11) 110°
12) 95°
13) 72°
14) 120°
15) 45°
16) 60°
17) 90°
18) 40°
19) 2
20) 9
21) 12
22) 5
23) 4
24) 7
25) 20
26) 35

27) 44
28) 27
29) 77
30) 14

Lesson 37
1) (4, 3)
2) (−1, −1)
3) (0, 0)
4) (0, 4)
5) (−5, 0)
6) (−3, −7)
7) $\left(-\dfrac{1}{2}, -\dfrac{5}{2}\right)$
8) $\left(\dfrac{15}{2}, -\dfrac{1}{2}\right)$
9) (2, 5)
10) (5, −4)
11) (3, −7)
12) (0, −6)
13) 5
14) 9
15) 10
16) 13
17) 5
18) 5
19) $\sqrt{41}$
20) $\sqrt{34}$
21) $p = -1$ or -9
22) $p = -1$ or 7

Lesson 38
1) $\overline{JK}$
2) $\angle C$
3) $\angle K$
4) $\overline{CD}$
5) $\angle N$
6) $\overline{KL}$
7) $\overline{DE}$
8) $\angle J$
9) similar
10) not similar
11) $y = 21$
12) $z = 12$
13) $z = \dfrac{15}{2}$
14) $a = 9$
15) 36 feet
16) 30 feet

Lesson 39

1)

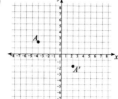

2)

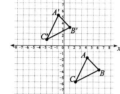

3)

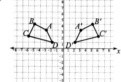

4)

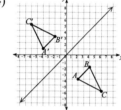

5)

6)

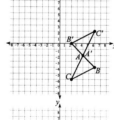

7)

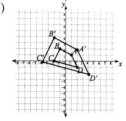

8)

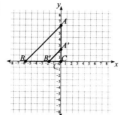

Lesson 40

1) 90
2) 110
3) 1,010,100
4) 6.5
5) 3.1
6) 17.1
7) 1000
8) 12,000
9) 9,032,000
10) 14
11) 1
12) 53
13) 48 ounces
14) 4.5 gallons
15) 1 pint
16) 4,300,000 pounds
17) 18 yards
18) 10,560 feet
19) 0.4 kilograms
20) 3,000,000 milligrams
21) 50 milliliters
22) 400,000 nanometers
23) 10 decagrams
24) 34 centiliters
25) 45.72 centimeters
26) 49.68 miles
27) 37.85 liters
28) 1.8288 meters
29) 946.24 milliliters
30) 121.275 pounds

Lesson 41

1) 5.44 kilograms
2) 24 square yards
3) 41.7 pounds per minute
4) 1×10^9 cubic millimeters
5) yes (She is 121.92 cm tall.)
6) 8.33 minutes
7) 144.81 kilometers
8) 402.34 meters

Lesson 42

1) A
2) A, B, C, D
3) E
4) A, C
5) C, E
6) B, E
7) B, F
8) A, D
9) C, D

Lesson 43

1) 34 cm
2) 18.84 in.
3) 12 m
4) 78 mm
5) 25,120 miles

6)

Rectangles

length	width	Perimeter
10 meters	**5 meters**	30 meters
14 feet	12 feet	**52 feet**
2 inches	8 inches	20 inches

7) $(16 + \sqrt{8})$ m or $(16 + 2\sqrt{2})$ m or 18.83 m
8) 48 inches

Lesson 44

1) 6.25 cm^2
2) 150 in.^{2}
3) 38.5 cm^2
4) 80.5 in.^{2}
5) 113.04 in.^{2}
6) 12 mm^2
7) 20 square units
8) 12 units
9) perimeter = 70 cm; area = 210 cm^2
10) perimeter = 26 in.; area = 34 in.^{2}

11)

Rectangles

length	width	Perimeter	Area
2	6	**16**	**12**
9	**5**	**28**	45
10	8	36	**80**
7	4	**22**	28

Lesson 45

1) rectangular prism; 8 vertices, 12 edges, 6 faces
2) square pyramid; 5 vertices, 8 edges, 5 faces
3) triangular prism; 6 vertices, 9 edges, 5 faces
4) cone; 1 vertex, 0 edges, 1 face
5) cylinder; 0 vertices, 0 edges, 2 faces
6) sphere; 0 vertices, 0 edges, 0 faces

7)

8)

9)

10) No three-dimensional figure possible; one potential "face" shares an edge with more than one polygon.

11)

12)

Lesson 46

1) L.A. = 144 ft.^{2}
2) L.A. = 47.1 cm^2
3) L.A. = 200.96 in.^{2}
4) L.A. = 384 ft.^{2}
5) S.A. = 251.2 in.^{2}
6) S.A. = 577.76 cm^2
7) S.A. = 760 ft.^{2}
8) S.A. = 12.56 in.^{2}
9) S.A. = 1296 in.^{2}

Lesson 47

1) 113.04 in.^{3}
2) 2612.48 m^3
3) 100.48 ft.^{3}
4) 56 cm^3
5) 360 m^3
6) 1728 in.^{3}
7) 77.94 cm^3

Unit 4 – Statistics and Probability

Lesson 48

1) positive
2) no correlation
3) no correlation
4) negative

5)

6)

stem	leaf
1	1 3 8
2	2 4 6 6 7 8
3	2 3 4 5
4	0 2 2 7
5	6

$1 \mid 1 = 11$

7)

stem	leaf
6	2 4 8 9
7	0 5 5 5 7 7 8 9 9
8	1 2 3 3 8 8 8
9	2 3 4 7

$6 \mid 2 = 620$

8) 38
9) 8
10) 62

Lesson 49

1) $y = \dfrac{1}{2}x + 4$

2) $y = -x + 8$

3) $y = 4$

4)

$y = 10x - 10$

5)

$y = -\dfrac{1}{2}x + 4$

Lesson 50

1) A: 162°, B: 36°, C: 54°, D: 90°, E: 18°
2) A: 25%, B: 16.7%, C: 8.3%, D: 50%
3) Item A: 25%, 90°
   Item B: 30%, 108°

Item C: 15%, 54°
Item D: 30%, 108°

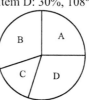

4) Item A: 24%, 86.4°
   Item B: 6%, 21.6°
   Item C: 40%, 144°
   Item D: 12%, 43.2°
   Item E: 18%, 64.8°

Lesson 51

1)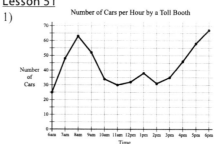

Number of Cars per Hour by a Toll Booth

a) 6am to 8am
b) 8am to 10am
c) Larger because the graph ends by continuing to trend upward.
d) No, it would likely drop off just like after 8am.

2) a) 10am to 11am
   b) 11am to noon
   c) noon to 2pm

3) 2002-2008

Lesson 52

1) a)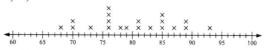

b) 25
c) 76
d) 68
e) 93
f) 75 to 85

2) a) {33, 34, 34, 36, 36, 36, 36, 38, 40, 42, 42, 42, 43, 47, 47, 49, 49, 51, 55, 57}
   b) 33
   c) 57

d) 22

e) 36

Lesson 53

1)

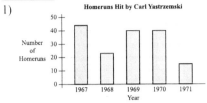

Homeruns Hit by Carl Yastrzemski

a) 162 homeruns

b) 73.9% increase

c) 25 homeruns

d) 27.2%

2)

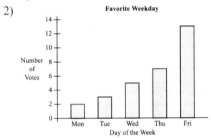

Favorite Weekday

3) a) 25 students

b) 12%

c) A line plot or a stem-and-leaf plot.

Lesson 54

1)

Interval	Frequency
11 – 15	2
16 – 20	4
21 – 25	5
26 – 30	4
31 – 35	3
36 – 40	2

2)

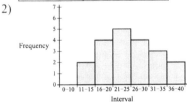

3) 30%

4) 21 to 25

5) a) 200 students

b) 68 students

c) 33 students

6)

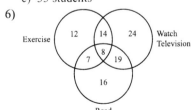

Lesson 55

1) mean = 43.8; median = 43; mode = 37

2) mean = 582.5; median = 595; no mode

3) mean = 26; median = 27; mode = 27

4) mean = 26.7; median = 28; mode = 18, 28

5) 23

6) the median (mean = 34.6; median = 35)

7) 48

8) 6

9) 88

10) 26

Lesson 56

1) 46

2) 43

3) 44

4) 75

5) 29.5

6) 58

7) range = 6; median = 5; LQ = 3; UQ = 6.5; IR = 3.5

8) range = 25; median = 23; LQ = 16; UQ = 29; IR = 13

9) range = 290; median = 230; LQ = 160; UQ = 350; IR = 190

10) range = 3800; median = 4100; LQ = 3200; UQ = 5000; IR = 1800

Lesson 57

1) 25

2) 32

3) 15

4) 21

5) 11

6) 21

7) 21

8) 50%

9) 75%

10) 25%

11)

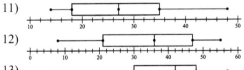

12)

13)

Lesson 58

1) C – Cluster Sample

2) B – Stratified Sample

3) A – Simple Random Sample

4) E – Convenience Sample

5) D – Systematic Sample

6) C – Cluster Sample

7) E – Convenience Sample

8) B – Stratified Sample
9) D – Systematic Sample
10) A – Simple Random Sample

Lesson 59

1) $\dfrac{1}{2}$ or 0.5 or 50%

2) $\dfrac{1}{2}$ or 0.5 or 50%

3) $\dfrac{2}{5}$ or 0.4 or 40%

4) $\dfrac{1}{5}$ or 0.2 or 20%

5) $\dfrac{1}{13}$ or 0.077 or 7.7%

6) $\dfrac{1}{7}$ or 0.143 or 14.3%

7) $\dfrac{4}{11}$ or 0.364 or 36.4%

8) $\dfrac{31}{50}$ or 0.62 or 62%

Lesson 60

1) 24
2) 36
3) 12
4) 1320
5) 48
6) 12
7) 800
8) 6,760,000

Lesson 61

1) $\dfrac{1}{12}$ or 0.083 or 8.3%

2) $\dfrac{1}{16}$ or 0.0625 or 6.25%

3) $\dfrac{1}{18}$ or 0.056 or 5.6%

4) $\dfrac{1}{6}$ or 0.167 or 16.7%

5) $\dfrac{1}{169}$ or 0.0059 or 0.59%

6) $\dfrac{3}{663}$ or 0.0045 or 0.45%

7) $\dfrac{1}{2}$ or 0.5 or 50%

8) $\dfrac{3}{8}$ or 0.375 or 37.5%

CPSIA information can be obtained at www.ICGtesting.com
Printed in the USA
269675BV00002B/1-24/P

9 780615 265094